CREATIVE TECHNOLOGY BRINGS FUTURE SOCIAL INFLUENCES

JOHN LOK

Contents

Preface

Introduction

Our global technology had been experiencing creative technology innovation stage. Creative technology will be experiencing to increase needs in global societies. They may include on medical creative technology industry, non-man driven or electric car transport industry, online movie or listen music or reading ebooks entertainment, even robot manufacture or service industries etc.

During to most of global industries need to own high creative technology to help them to improve service performance or or raise efficiencies. Hence, future creative technology industry social needs will be influenced to increase and to assist any kinds of industries development.

In my this book, I shall give evidences to explain why and how what kinds of industries , which will increase creative technology industry need. Readers can bring fresh creative industries development knowledge.

Prologue

Table of content

ONE

CREATIVE NON-MANUAL DRIVING VEHICLE TECHNOLOGY INFLUENCES

What the psychological need differences between rail and bus passengers

- Reasons we need to improve public bus transport tool service quality

The ways that we need to improve public transport, e.g. bus transport service, we try our best to ask these questions: During periods of stress on the bus, like weather conditions or maintenance failure that slows the bus service system? How to improve mass transit on bus service frequency, when looking at ways to improve public bus service transport , riders want frequency? Interestingly, speed is not as much of an issue, if they are waiting downtown in the rain, or on some suburban backstreet, riders want to know that a bus will arrive soon, preferably in less than 15 minutes. Therefore, the wait becomes part of the transportation cycle. Even, if the bus is lightning fast, in the mind of the rider, the trip begins right when they arrive at the bus station, and start waiting for the bus to pick them up.

`What does efficient bus ticketing system mean? It is big part of how to improve bus transportation efficiency is improving transit ticketing system, because ticketing systems have to be quick and practical to allow for prompt loading and unloading of passengers. So, inefficient ticketing systems also slow down bus frequency, as drivers need to wait for everyone to tap before they can drive away to the next stop.

How to let passengers feel comfortable? Riders want comfortable buses that can seat as many people as possible. Face-to-face seating is not appealing and being knee-to-knee in a confined space creates awkward moments between strangers. However, comfort also extends beyond the buses' seating arrangements. A smooth riding, quiet bus plays a significant role in reducing the overall stress of a public transit experience. Among the consistent feedback from riders of fuel cell electric buses is a surprised delight about how quiet the buses are when in motion.

On reduce greenhouse gases environment prote3ctoin aspect, exhaust spewing buses are on ongoing concern. One of the significant factors that commuters consider when deciding to take public transit is the environment impact of their alternative transport method. And although a diesel bus packed with 40 people may be less environmentally damaging than 40 separate diesel cars, it will still have negative impacts on both local air quality and the overall climate situation , when given the choice, we've found nearly all riders prefer " zero-emission buses" to conventional diesel buses nowadays.

IN fact, we are always thinking of ways to improve public transportation by dev4eloping new clean fuel technologies. Fuel cell electric buses resolve some of the above issues for both transit bus operators, bus performance is continually being proven and improved over millions of miles of operation in environments ranging from mountain villages to desert communities to busy cities. Hence, the first step to creating better public transit networks is becoming aware of the available options. Many communities are taking measures to improve public transport by implementing innovative sustainable transport solutions that have profound impacts on the live ability of their communities.

So, I shall recommend these ways to improve public transport methods to bus service as below:

Firstly, making interchanging easy for public transport has most efficient public transport service improvement aim at linking areas that are outside a city to the city center., doing this is beneficial in two ways. It helps people who should not at the city center , but needed to pass through because the outlying areas are not connected together to keep off and hence reduce congestion at the center. Also, connecting the outlying areas provide a backup for the public transport system in case of a problem which often happen.

Secondly, minimize the number of stops/ stations, stops and stations improve the efficiency of public transport , but there should be a balance between enabling accessibility with more steps or stations and reducing the costs of operation by increasing transit need of ensure trips are covered in time. Therefore, core should be taken to ensure that stops and stations are located on streets to balance accessibility by commuters on one hand and reduces operating cost on the other hand.

Thirdly, lessen traffic congestion by deploying a number measures. Reducing traffic congestion at city streets could be done, implementing a number of strategies, such as providing lanes dedicated specially for the use of public transport, deploying strict regulations , such as queue bypasses or queue jumps. Another means of reducing traffic congestion is by providing feeds and data from public transport systems, freely to commuters to educate and help them avoid areas of traffic congestion and finally, giving priority to public and trams operating efficiency, increasing the travel time of these engineering mechanism whereby a traffic signal turns green at the light of a public transport at an intersection. All of above these improvements may be future public transport bus passengers service improvement need, if any bus companies hope to increase their bus passengers number absolutely.

- What rail passengers really want rail innovation improvement

Public transport systems, such as rail provides benefits including less traffic congestion, less pollution, safe travels, lower expenditures , less effort and better predictability in comparison to road transport. In fact, bus and train riders experience the most negative emotions in comparison with other transport modes, such as private cars , walking and cycling. Hence, technology has the potential to bring about the changes, needed to increase efficiency of rail transport, e.g. cost-effective ways to improve the quality of public transport and increase ridership may involve comfort and convenience improvement, or technology has the potential to provide more up-to-date information and customized service to train passengers and therefore improve the rail journey experience . On the overall, passenger journey , e.g. the importance of automated traveller information systems, and electronic fare payment collection systems can bring rail passengers look for this information in different interfaces from localized displays installed on platforms to smartphone applications.

Moreover, technology can also improve fare collection and management which of made manually can be prone to error, and time consuming , unified cards, smartphones can make it easier for rail passengers to obtain ticket, with the potential to increase the user satisfaction with the rail system. Because rail passengers demand not only pre-trip information for planning their travels, but also information during journeys, such as punctuality, connections and platform allocation. One extensive review indicates that accurate communication, for example, giving effective way finding information, can optimize passengers' experience with public transport.

Also, technology can facilitate the process of finding free seats on trains, which is a current demand from rail passengers and the cause of stress during the boarding process. IN fact, many rail passengers have specific preferences regarding seats and would appreciate having control of where to sit. So, navigation and way finding information can be delivered directly to passengers to inform where they could stand aiming to board less busy carriages, for example, choosing to travel on a less crowded train, or spreading themselves out on the platform before boarding in respond to crowding information, e.g. smartphones are frequently used by passengers of public transport and can make waiting times seem shorter. Furthermore specific system features designed for train passengers have the potential to improve the journey experience of the travelling public.

What ferry passengers service improvement need

- How can ferry service be improved affordable, reliable, convenient, flexible and clean will get drivers out of their cars ad onto environmentally responsible to passenger ferries?

Ferry transportation provides an environmentally friendly commuting alternative to the congested roadways in many of countries , so ferry transport service needs to meet long term air quality goals, it is critical to move beyond traditional technologies to zero-and near zero emissions technology. Clearly putting a transit system in operation that demonstrates emission control technology and the development of zero-emissions, ferries will help achieve air quality goals to our societies, for example., new shipping rout4es are needed to increase in order to satisfy ferry passengers different rapid ferry journey short distance need, when they need to choose one kind of public transport service either bus or rail or ferry transport service among of them.

None ferry accident occurrence, ferry service needs to let passengers to feel it is the safest sea pubic transit, expanded recreational service is also needs, particularly on weekends when bridge , corridor traffic congestion is becoming an increasing problem. Ferry service needs have uniquely provided flexible, vital transportation supports in response to a natural or man-made disaster that shuts down bridges and roads, fuel –cell technology is needed , that will lead to zero-emissions ferries, e.g. on-board emissions monitoring is far less polluting than previously through, e.g. 149 passenger boats are designed to travel 25 knots or less , and 300-350 passenger vessels designed for speeds up to 30-35 knots.

This emissions standard will perform specifications and the cost of this technology is accounted for in the ferry company vessel capital budget ,e.g. vessel design capabilities to accommodate existing and new docking configurations . This maximizes fast ferry passenger loading, including bicycles, carriages and wheelchairs. Hence, future global ferry service needs have these positive influence to our societies: Need for flexibility, desire to help the environment, need for time saving, which includes the importance of reliability, sensitivity to personal travel experience, such as a need for personal space or quiet feeling ferry seat any time, insensitivity to transport cost, e.g. the ferry ticket price is cheaper than rail or bus fares sensitivity to stress.

However, ferry service is different unlike rail, bus because expanded ferry service can be launched quickly at low initial cost and with great flexibility. Unlike buses, ferries are not hindered by traffic congestion on roads and highways or in tunnels. So, ferry service can be safely expanded to bring new service to new places and add more service to existing routes more easily than bus and rail public transport both, e.g. expanded ferry transport service can operate safety and provide with a robust, flexible and effective emergency response capability if the region is hit with a natural or man-made event that disables roads, other transit, bridges , before any.

Hence, ferry companies need to decide to improve their ferry transport service, they need to answer these questions: Is the new shipping route a good transportation investment? Does the new shipping route have fatal environmental negative impact? Does it offer a transit option that can be initiated in a timely and cost-effective manner? Can it provide ferry transport service that is reliable, safe and fully accessible after the ferry recovery would be unreasonably high charge to ferry selection is decided to implement to increase?

Also, ferry safety is needed to consider because it can influence any ferry passenger choice, when the ferry is moving on the sea, when the passenger is sitting on the boat. The ferry safety issue may include: Ensuring that access to all ferry operational areas, including, machinery spaces, pilothouse and gear lockers, remain locked at all times and accessible only to authorized crew, posting night watch security guards at terminals, conducting diligent onboard inspection for unattended passenger bags, briefcases and packages after each run, before the next boat load is allowed to board, creating coded signals and response to report suspicious activity, requiring positive identification before allowing any contractors, vendors or others access to ferries, providing additional security training to crew, developing a security plan to account for potential threats, outlining preventive measures and detailing an action plan in the event of a threat or actual emergency.

Future Human Transport Need Change

How future our transport need change? What factors influence our future transport need change? In general, these factors may influence our transportation need change. They may include fuel cost, the labor market for commercial drivers, demand for frieight , customer loyalty , vehicle capacity, government regulation, geographical events, the public transport tool reputation to passegners as a merchant. However, the factors that influence the development of transport system in an area? They may include as below:

Environment at the local scale existing hydrographical and geomorphological characteristics are string, factors in transport development, particularly in terms of the technical challenges (bridge, gradients,) they present to construct, other factors may include historical, technological, political and economic factors. All of these factors may influence our future transport system how develops. For raiway development influential factors, they may include: Geograohical factors, e.g. the North Indian plain with its level land, high density of population and rich agriculture presents the most favourable conditions for the development of railways in India. However, the presence of large number of rivers makes it necessary to construct bridges which involve heavy expenditure to Indian Government publich transport expenditure.

How transport has changed from past to present?

There has been a remarkable development in modern transportation. The stream engine and then the stream trains have emerged and spread at this time and in abundance until the discovery of natural gas and oil was an evolution of transportation. Thus, the sedams and vehicles began to run in oil, until present battery changes energy vehicle need, even future non-manual driving artificial intelligent driving vehicle need. These new transport technology may influence our future public transportation from gas energy to battery changed energy, even non-manual driving vehicles need to our daily transport need.

So, our future purpose of public transport need is the unique purpose to oversome space, which is shaped by a variety of human and physical constraints, such as distance, time. These both is our future main public transport need main purpose factors, short distance and reducing journey time, they influence that why we need to choose to catch any kinds of public transportation tool to replace purchase private cars to drive transport tool choice. So, future any kinds of public transport tools, they need to consider above both main factors , how to attract passengers to choose to catch themselves public transport tools choice in this competitive public transport tools market.

On the other hand, the economic importance of transportation development can be defined as improving the welfare of a society, through appropriate social, political and economic conditions , such as US Government spent too much money to assist MTR (MAss transport railway firm) to develop underground thrain transport. Its aim to let many passegner can reduce journey time and reduce distance between destinations, it also hopes US citizen passengers can pay cheap transport fare to buy ticket to catch underground transport train for many families their transport expenditure in social transport welfare view.

However, US Government neds to solve those challenges, before it implements to develop rapid underground railway , e.g. lack of knowledge of geographical fwatures, lack of manpower necessary to operate the rapid underground railway construction work, lack of construction materials within the US itself. For Brazil rail network transportation development example, the factors influence the use of rail network for transportion is highly restricted in Brazil. Thus, the development of roadways and waterways is the main modes of transportation that caould be used in Brazil given its topography and drainage benefit to society . So, brazil can develop rail network for transportation development in success.

So, transportation system is important in the development of any nation, because transportation plays important role in rapid economic growth of a nation. Thrapsortation increases the quality and variety of consumer goods, thereby stimulating the demand and development of trade and economy of the nation. Moreover, transport provides various employment opportunities and boosts up the economy of the country.

Also, any transport tools need to improve themselves transport service in order to attract passengers to choose their public transport service more easily. They may attempt to sign up for an autonomous vehicle pilot program, free phone enquiey concerns whether the passegner can catch which bus bumber to go to the destination, hou much bus fare, how long journey time, when the bus will arrive teh bus stops or leave the bus stop etc. bus service questions, before any one passenger prepares to choose to catch bus (free bus go phone call enquiry), free download a public transport tool transit app. even water taxi tranport tool innovation can replace ferry public transport tool, it can let passengers have more fun an enjoyable catching feeling. So, water taxi tranport tool is one kind of future new transport tool change to replace ferry , it can influence ferry passengers to choose water taxi public transport tool to replace ferry. Although, its fare may be more expsnse to compare ferry, but it can reduce jounrey time and distance between both water stations, when ferry can not arrive the other destinations, but water taxi can arrive any one water station destination. It can bring convenient to future any one ferry passengers. So, water taxi may be developed to some countries, e.g. New Zealand , Auckland city, US , Washington and New York cities they had developed water taxi public transport tools to let ferry passengers have one kind new water public transport choice.

However, instead of new transport innovation improvement to water transport service public transport with input from the public on bus transport service aspect, bus frequency improvement, it means when booking at ways to improve, bus frequency from long times to less times, efficient bus ticketing system, a big part of how to improve tranportation efficiency is improving transit ticketing system.

In fact, my future transport system may still include these five types, modes of transport are: railway, roadways, airways, waterways and piplelines. Also, among different includes of transport, railways are the different modes of transport, railways are the cheapest. Trains cover the distance in less time and comparatively, the fare is also less to other modes of transporation. Therefore, railways is the cheapest mode of transportation to compare ferry, water taxi , sea transport, bus, taxi, road system.

On conclusion, transport price is not the main factor to attract passegners to choose to catch. The importance to have a good public transport system in place. It may be one main factor to help the kind of public transport tool to attract passengers to choose to catch, because a good transport links can widen people's job search area and help them find employment. It can also reduce commuting times and reduce the cost of living, and high skilled workers are more likely to travel across longer distances to work, especially if they are following good job opportunities. So, future any one kind of public transportation tool service provider ought consider how to satisfy working people working time need to shorten journey time to any working places or student learning time need to shorten jounrey times to any schools as well as let they feel comfortable to sit on comfortable chairs or provide free internet service to themselves mobiles , laptops, when they are sitting down or standing up in the kind of public transport . It is the important factor to influence any kind of public transport service in success.

Future Non-Manual driving vehicle How
Influences Public Transport Tool Passenger Need

Nowadays, artifical intelligent (non-manual) driving vehicles are invented, it may be accepted to any countries families to feel comfortable to drive on roads, because any people choose to buy any kinds cars, when any people choose to buy kinds of non-manual (artificial intelligent) vehicles, they do not need to use their hands to drive cars, because artificial intelligent (robotic auto control wheels, it means that robots can help human (drivers) to control wheel to drive to avoid any cars crash occurrence on the roads more easily.

If one day, non-manual driving robotic control whoole vehicles are invented in successful, whether it will persuade many different conuntries families choose to buy non-manual (robotic auto control wheel) vehicles, then it will cause bus, tram, train, underground train, road transport need will be influenced to reduce or even if non-manula boats are invented, whether it will cause ferry sea transport needs will b influenced to reduce. Hence, future non-manual driving vehicles or bats invention whether they will influence public transport tool of road and sea transport passengers number reduces. It is one interesting question. I shall attempt to discuss as below:

In fact, non-manual vehicles are very attraction, to excite any person chooses to buy to drive, because people do not need often touch wheels and touch foots button to control cars to move often forever, when robotic can be invented to help human to control car wheel and foot button, any person only needs to sit on his/her car, then the car can move rapidly, because any drivers is lazy, he/she hopes machine can help her/him to drive car on the road safely. So, he/she can read book or listen music or eatch mobile movie to enjoy his/her entertainment when he/she is sitting on his/her car.He/she will feel more comfortable and enjoyable when robotic can help him/her to drive car. So, robotic (non -manual driving vehicle) can encourage people to choose to buy cars because any drivers won't need to drive cars, robotic can help drivers them to drive on the road easily, when global any one family can own one robotic auto control (non-manual driving) car at least, it may influence these owning non-manula diriving vehicle owners do not feel need to pay any fares to buy road public transport tools of bus ticket, train ticket, underground train ticket , tram ticket to go to anywhere. So, it seems that robotic (non-manual driving) vehicles may influence future any road transport passengers number reduces , because traditional catching any kinds of road public transport tool passengers will be influenced to choose to sit themselves auto (non-manual) driving cars to go to offices to work, parents do not need to follow their sone/daughters to sit on themselves non-manual auto driving cars to go to schools, because their sons/daughters can sit on themselves non-manual driving cars to go to schools more easily. In holidays, they can sit on themselves non-manual driving cars to go to cinemas, music halls, breachs, theaters, shopping centers, gardens different entertainment places to enjoy their any leisure safely because robotic can help them to drive their cars on roads safely.

So, it means that robotic auto control driving cars can influence global every family to feel that they do not need to catch any kinds of public transport tools, e.g. bus, train, tram, taxi underground train to go to anywhere because robotic auto driving cars can help any one, he/she does not know how to drive car to go to anywhere safely. So, future any one won't need to learn driving car skill, when he/she likes to buy one auto driving car. So, in passenger public transport need view, non-manual driving cars will influence them to feel any kinds of road public transport tools can help them to go to anywhere conveniently, because themselves non-manual driving vehicles can help them to drive cars to go to anywhere conveniently. They only need to tell robotic that where they want to go, when they sit on their non-manual driving cars, then robotic knows whether where destination, they want to go, their cars will auto move on the road immediately. It is one exciting and enjoyable ourney when the driver does not need to drive his/her car on the road. So, it seems that robotic (non-manual driving) vehicles invention may bring negative influence to any kinds of public transport tools service needs to passengers , when passengers had owned one non-manual driving car at least.

Why and how non-manual driving car owners need

raise public transport quality on travel time and fare

aspects

● How non human driving behavior can be influence by non-manual driving cars

In fact, impact of automated vehicless on travel mode preference, it can bring both trip purposes and distances aim raising need to any kinds of public transport service passegners. Because of technology penetration in the transportation system, the automated vehicle is set to be a future mode of transport, it may bring negative impact to future any kinds of public transport passengers needs, in special on the potential impact of these non-manual driving automated vehicles on travel behaior negative impact to public transport passenger behavior. Automated vehicles will influence future public transportation passengers feel it can bring more short time travel distances and short trip purposes more benefit than any kinds of public transport choices, e.g. bus, taxi, ferry, train, tram, underground tram etc. road and sea public transport tools, e.g. ferry, water taxi. It means that when future any passenger feels above these any one kind of public transport tool needs to spend longer travel time on journey distance and trip to compare future automated vehicles, then they will choose to sit on automated vehicles in preference, due to automated vehicles can help global any one person needs to go to anywhere rapidly.

So, automated vehicles may replace general traditional public transport tools in possible, when they are popular accepted in societies. On the other, instead of shortening journey travel distance time, (travel time) aspect, public transport fare, travel cost will be another influential factor to influence future public transport tool passengers to choose automated vehicles to replace to catch any kinds of public transport tools.

In fact, conventional cars and public transport s are perceivd as being the least attractive alternative in relation to in-vehicle travel time on short and long distance communting trips. So , future automated vehicle drivers (non -human driving) behaviors will be likely changed to prefer this mode for long distance leisure trips rather than short distance commuting trips by automated vehicles.

In fact, advanced technologies have revolutionized many aspects of human life, include the automated vehicle transport system. Also, transport system is one of the essential development aspect to particular , such as non-manual driving automation , vehicle aims to make trips safer, faster , more efficient, automated vehicles passengers and drivers can feel enjoyable to do themselves leisure behavior , e.g. read books, listen, music, listen mobile, watch laptop movies when any one does not need to consider whether their cars are safe to be driven , even any one needs to drive the automated car, because robotic can help them to control how to automatic drive this car on the road safely.

Robotic will bring confidence to let them feel that themselves cars are moving safely on the roads . In recent years, the concept of automated driving has been introduced as on outstanding platform for the next generation of driving systems that is expected to improve safety, traffic flows efficiency, reducing traffic jams occurrence chance, avoiding traffic accidents occurrence chance, e.g. avoid to crash any one person when he/ she is walking across road or crach any car is moving on the road easily, capacity, accessibility , and reducing congestion through the application of some technologies , such as vehicle to vehicle and vehicle to infrastructure communication.

So, future automated vechicles can have good driving facility systems to be installed in their cars, in order to raise safety, rapid driving speed level to let any one to feel , when they are sitting in their automated cars, e.g. using cameras, sensors, global positioning system adaptive cruise control, light detection and ranging, and advanced driver assistance system, automated vehicles can steer the vehicle and drive it automatically when passengers delegate control to a computer. Absolutely, ny replacing the driver role with an automated driving system , future one automated vehicle is able to totally free up passengers under automation levels.

So, unless future any kinds of public transport tools may apply automated robotic automated driven system replace the bus driver, taxi driver, train driver, tram driver, underground train driver to raise automated driving system service improvement level to let any one passengers to feel. Otherwise, when automated vehicles are popular to be accepted to buy in any one country in global. Then, global public tansport tool passegners number may be influenced to reduce when global any one family owns at least one automated vehicle at themselves homes .

In other words, automated vehicles can bring thes benefits to let global any one household family feels, future automated vehicles users , they can mostly behave like passengers inside the vehicle, which implies that they will be able to multitask and productive by allocating the travel time to do other activities, e.g. reading, eating, working, drinking, watching movies, listening musics, even sleeping. So, automated vechicles will motivate humans to change non-humanly driven behaviors from conventional humanly driven behavior. This non-humanly driven behavior may be one main factor to influence or encourage future any one kind of public transport passenger won't choose to pay fare to buy ticket to catch any one kind of public transport tool again, because non-manual driven behavior may hel many lazy people do not need to consdierate how to learn to drive cars skills to prepare pass any road test in order to earn the driving licnece to permit to drive cars forever. When automated vehiclesa re popular to be accepted to replace manual-driven cars in societies.

Hence, automated vehicles could potentially change the traditional human driven vehicle market to cause their manual driven cars sale buyers number reduces, when the automated vehicle buyers number increases, also they can chance globa public transport passengers behaviors to reduce to pay fares to catch any kinds of public transport tools when automated vechicles users may sit on themselves automated vehicles to go to anywhere in short time rapidly and safely in any countries.

On conclusion, future global public transport service competition is serious, because instead of global passengers had began to compare whether which kinds of public transport fares are cheaper, more safe, shortening journey time between leaving place and destination, more comfortable feeling, e.g. clean and comfortable chairs , mre free internet service facilities in order to make any one kind of catching public transport tool choice in preference. On the other hand, future automated vehicles number will increase when traditional manual driven car users begin to believe that automated vehicles can bring more safe , more comfortable, more fee- time using, more leisure satisfactory feeling, more than traditional manual driving cars. Then, when global any one household family had made choice to buy at least one automated vehice to replace themselves car(s) at home. When, they are habit to sit in themselves automated vehicles to go to anywhere, however, short or long trip . Consequently, global any one household family won't feel any kinds of public transport tools may bring personal economic saving cost, comfortable, enjoyable, free-time using benefit to compare themselves automated vehicles . It will cause global public transport tools passengers number will reduce , when many different kinds of home automatic vehicles are purchased to replace manual driving cars by global household automated vehicle users. So, in passegner transport tool choice psychological view, automatic vehicles will be possible to replace future public transport service tools. So, any public transport service providers can not neglect how to desing and improve their facilities , charge reasonable transport fare, provide more comfortable, and enjoyable sitting feeling , even applying automatic driving system to replace human drivers in order to attract passegners ' catching need choice more easily.

Artificial Intelligent In Road Transportation Strategy

- How artificial intelligent vehicle may interact intelligent transportation tools

Can artificial intelligence (AI) and machine learning (ML) be used in the search for new " consumption" behavioral type variables that affect consumer individual or transportation service organization individual different transportation tools choices, such as road or sea or sky transportation tools? Can artificial intelligent vehicle may interact intelligent transportation tools market development?

Consumers usually have bargaining and on risk choice when they are already shopping, such as who need to accept to use any (AI) new technological products to replace human traditional behaviors, such as intelligent non-manual driving transportation market, e.g. cars are needed to be driven by human drivers on road, but it has bargaining and on risky choice, when non-manual (AI) vehicle buyers who need to depend on non-manual artificial intelligent (ML) system assists them to drive their cars on the roads.

So, any non-manual driving auto car buyers must need to believe (AI) non-manual driving vehicles (ML) systems can make accurate driving judgement to reduce or avoid any traffic accident occurrences more than human drivers' driving judgement when the (ML) systems are driving their cars on the roads. Then the intelligent vehicle manufacturers will have possible to sell their non-manual driving vehicles success.

This is the first reason or idea influences consumer individual choice to buy any kinds of (AI) non-manual driving vehicles, when consumers believe (ML) systems are more safe and make more accurate judgement to compare human or computer systems, when they are sitting in one non-manual auto driving vehicle on the road.

The another second reason or idea is that some common limits on driving consumer prediction might be understood as the kinds of errors made by poor implementation of machine learning.

Supposing driving consumers believe (AI) machine learning ability is worse to compare to human learning ability. It will also influence driving consumers do not accept to use any (AI) non-manual auto driving vehicles to replace every driver is essential on driving by himself/herself on the road. The third idea or reason is that it is important to influence driving customers believe how (AI) non-manual auto driving technology is used in them can both overcome and exploit human driving skill and safe limits and raise more auto driving safe judgement to compare human driving safe judgement.

However, how to predict any kinds of (AI) non-manual driving vehicles future consumption effort, due to different kinds of (AI) non-manual driving transportation vehicles which have different unique functions and designs to be used by different kinds of road transportation or driving demand of consumers. For example, lorry drivers need non-manual intelligent system can help them to drive fast, but safe to assist them to transport cargo to arrive destinations from their factories or offices. Otherwise, private car driver expects whose (AI) non-manual driving vehicle can auto drive to send to whom to arrive destination in safe way and non-too fast and non-too slow speed in order to avoid accident occurrences.

So, a different road intelligent consumer demand is to define whose individual driving behavior and driving habit and driving attitude and driving judgement and driving speed demand to decide how to design whose intelligent vehicle to satisfy those driving demand more generally, as simply being open-minded about what variables are likely to influence every consumer economic choice, when who decide either to buy any kinds of (AI) products or not to buy any kinds of (AI) products to replace the different demand of consumers their different (AI) useful demand.

Hence, for these three (AI) products group of stakeholders, such as home (AI) consumer group, firm (AI) consumer group and government (AI) consumer group . These consumer groups may consider whether different kinds of (AI) products can give what is special beneficial interest to them to use. These variables can be measurable properties of choices to influence them to choose to buy any (AI) kinds of (AI) products to use, e.g. psychophysiological, biological, social influences, consumer's wealth, moods and personality, (AI) product price etc. variable factors which will influence them to decide to attempt to buy any kinds of (AI) products to use. If behavioral economics is as open-mindedness about what variables might predict. Then , (AI) machine learning system is a way to do behavioral economics because it can make use of a wide set of variables and select-which ones predict.

In behavioral economic view point, when general consumer overall demand to the product is much than the other similar (AI) non auto driving vehicle products, such as any kinds of (AI) non-manual auto driving vehicles and any kinds of manual driving vehicles case, then any kinds of (AI) non-manual auto driving vehicles will be more attractive to cause many manual driving vehicle buyers choose to buy (AI) non-manual auto driving vehicles. Hence, it seems if any kinds of (AI) non-manual auto driving vehicle products can make more attractive variable efforts to influence overall driving consumers to feel that they have more needs to drive non-manual auto vehicles to compare more than driving manual driving vehicle.

What is the main variable effort to intelligent vehicles to attract driving consumers to choose to accept to drive them ? However, I believe that (AI) machine learning system is a main factor to raise overall driving consumers' acceptances to drive it to replace manual driving vehicle. If it can persuade or prove (AI) machine learning system ability and judgement effort is more accurate than human or computer learning effort or judgement effort, then it is possible that any kinds of (AI) non-manual driving vehicle products will be accepted to drive on the road in popular.

Machine learning system is able to find prediction value in details of how the bargaining occurs. This discovery is the beginning of the next step for driving consumer individual driving behaviors or driving habits. It raises questions that include: What variables predict to influence driving consumers to change whose driving habits or driving attitudes? How can driving consumer individual emotion, face-to-face talking with whose friends when they are sitting in the non-manual driving vehicle to influence whom driving habit or driving attitude to be changed ? Do driving consumers consciously understand why those habit driving attitudes variables are important when they are sitting in one intelligent vehicle? Can (AI) driving machine learning methods capture the effects of motivated cognition to influence driving consumers decide to buy any kinds of (AI) non-manual auto vehicle products more attractively. So, it seems (AI) driving machine learning method is a main variable factor to influence driving consumers to feel who have more confidence to drive them more than any other kinds of similar manual driving vehicles on the road.

Consequently, (AI) driving machine learning system will be one important psychological method to influence driving consumers to choose to buy (AI) auto driving vehicle products to replace manual driving vehicles. The reason is because human and driving machine learning system both which will have limited variable factors to influence general different countries (AI) driving consumers' need desire to be raised.

- Why can (AI) driving machine learning system main factor influence driving consumer individual desires ?

Driving consumer expectations are hard to measure or predict driving attitudes and driving behaviors in (AI) non-manual driving vehicles market. Artificial intelligence is another kind of computer science development to apply intelligent vehicle market. Why do driving consumers feel need to buy any kinds of (AI) auto driving vehicles to drive to replace manual driving vehicles on the roads? What are (AI) auto driving features different to manual driving features?

(AI) is the recreation of cognitive functions in computers; it enables machines to perform tasks like humans and perhaps even better than human. In the real world, scientists develop the technological singularity, in which a superintelligence emerges with unfold human consequences.

Professionals in many industries are intensely interested in the specifics of what (AI) can do today, and how can it helps. They are considering the impact of applied (AI), in which computers are used to address a particular problem, extracting and utilizing patterns found in large volumes of data. Of all (AI)'s subfields, machine learning is attracting the most attention. I shall explain why (AI) machine learning system is the main factor to lead consumers feel need to buy any (AI) products to use. Such as below:

For smartphone, fraud detection to medical diagnosis etc. applied (AI) technological products examples. (AI) machine learning systems can help any one of these products to do any exceed general computer learning systems which (AI) learning systems can do any skills to supply (AI) users to use to compare computer learning systems can not do any skills to supply compute users to use. It seems that (AI) machine learning system is the unique feature to attract consumer consideration in technological product market.

An term for different types of learning, and can be accomplished using different techniques. This has led to a perception that all marketing teams should have (AI) to bring a unified personalized customer experience, when consumers choose to buy any (AI) products to feel what are the different or unique characteristics to compare general computer products. Such as (AI) product has this unique machine learning characteristics, we can predict (AI) and machine learning is connected to influence consumers to feel needs.

Furthermore, over the same time period, and in contrast to predictions for roles in many industries. (AI) won't take the place of marketers and merchandisers themselves although it is already a new value to analytical and strategic marketing skills to persuade consumers to buy any (AI) products. It means different kinds of (AI) products will have different machine learning effort and unique characteristics to attract consumers to choose to buy them to use. Such as, when intelligent vehicles need have unique road driving or sea transportation or flying machine learning system when they are applied on these three kinds of transportation tool aspects. They need have good response safety driving and immediate response learning systems to avoid any boats or air planes or vehicles to crash to them to reduce accident occurrences immediately on any one of either road or sky or sea journey environment.

What is the reason why (AI) driving machine learning system can influence good at making sense to driving consumer desire? Only humans (drivers) , preferably experienced, well informed humans can understand their driving customer needs and decide how to design or reengineer any (AI) intelligent vehicle product functions. (AI) intelligent vehicle can give these professionals the means to do this better to compare manual driving immediate response control function when any vehicles are driving or they will stop immediately to close / near to them in order to reduce crash occurrence on the road, and then maximize relevance through real-time customization of the non-manual auto vehicle driving user experience.

For example, as ever, senior decision makers need to be informed, decisive and results-oriented or risk losing out. Harvard Business Review indicated : Over the next decade, (AI) won't replace managers, but managers who use (AI) will replace those who don't. Such as intelligent vehicle won't replace drivers, but drivers who use intelligent vehicles will replace those who can not control how to drive their vehicles in the most safe way. So, (AI) driving machine learning system will have possible to do any drivers' (human's) driving judgement, driving analytical mind and driving effort to be more accurate than manual driving skills. Such as how to control to drive the intelligent vehicle in the most safe way. It is general manual driving skill can not achieve to drive in the safe way.

For another (AI) digital commerce example, (AI) and machine learning are the most exciting developments in marketing and merchandising to be applied to digital commerce, such as making better decisions through trend and cluster analysis, deploying product and content in mutually reinforcing combinations, increasing customer engagement and satisfaction in real time.

Hence, the key attraction in digital commerce circles is that machine learning is designed to be self-optimizing. Optimizing for revenue example will surface are increasingly profitably selection of products (within the brand parameters selected).

When to apply (AI) capabilities and what value (AI) is delivering for customer and company like. Unlike any technology before it, (AI) is analytical and predictive capabilities offers the prospect for each and every individual. It can maximize real time and engagement. Effective tailored (AI) technology, such as digital experience cloud technology is available now. And once integrated, (AI) starts learning and delivering incremental value from day one. So (AI) could transform the digital experience to any business organizations.

Hence, (AI) driving machine learning system can be applied to road driving skill aspect. When intelligent vehicles are invented to own the most safe driving judgement skill and they can know when either they may auto drive fast speed, when they are feeling to know when there are not many vehicles are moving close/near to them or when they need auto drive slow speed, when they are feeling to know when there are many vehicles are moving close/ near to them. Then driving consumers will have more confidence to choose to buy any kinds of intelligent vehicles to replace manual driving vehicles to drive on the roads.

- Non-manual driving transportation tool market development

If Non-manual driving vehicle manufacturers expect their (AI) automatic vehicles can attract drivers to buy. I feel them to need to consider how (AI) driving machine learning system can achieve these requirements in order to satisfy manual driving vehicle drivers' requirement to change their traditional driving habit to choose non-manual driving needs. It means (AI) driving machine learning systems can help them to drive vehicles to replace manual driving vehicles on the road. This is the main factor to influence car buyers choose to buy intelligence driving vehicles replace to manual driving vehicles. I believe (AI) non-manual driving vehicle machine learning systems, need to be designed as below:

(1) Improving driving safety by preventing accidents from happening.
Every year, drivers are facing a large number of casualties, due to traffic accidents. The amount of killed and injured road traffic related accidents is increasing every year. The real cost of an accident can go well beyond the limits of immediate material destruction, and is impossible to evaluate.
Hence, researchers and car manufacturers are looking for solutions in order to reduce the amount of accidents. They already developed a considerable set of technologies in order to decrease the amount of casualties. Most of them (like airbags, seat-belts, anti-lock systems, shock absorbing car bodies) are efficient in decreasing the impact of an accident, and in protecting the passengers of the cars. The technologies already saved a lot of lives, but they are rarely able to avoid accidents because they do not anticipate them. Moreover, if they are protecting in many cases, the passengers of the car, they do not prevent most traffic participants, like pedestrians on bicyclists from getting injured. it causes (AI) non-manual automatic car manufacturers need to consider how to design machine learning safety system is to prevent accident from happening instead of just reducing their impact.

This can only be possible using intelligent systems that can observe the driving environment, reason and decide if there is a danger, determine how to avoid it and act if necessary

(2) Reducing energy consumption by optimizing the driving.

Nowadays, global air pollution is serious. (AI) non-manual driving car manufacturers need to concern how to design (AI) machine learning system can reduce degree of air pollution to be the most minimum level to compare to traditional manual driving vehicles.

The reduction of energy consumption if certainly one of the main challenges. Transportation is one of the major factors in fossil energy consumption, and it is also responsible for a large amount of CO2 pollution. It is difficult to ask individuals to voluntarily limit the use of their vehicle of they do not have a strong incentive to do so. Specially in regions where vehicles are needed to drive to go to work every day. It stands to reason that if it is difficult to decrease the amount of vehicles, part of the solution is to make them more energy efficient.

Hence, non-manual driving car manufacturers need to design how to improve engines, which are more optimized and need less fuel to operate, and hybrid and electric cars have been developed and are continuously being improved. But we can go beyond these solutions that do not take into account the environment in which a vehicle is driving. A growing number of scientific contributions presented intelligent systems used in order to improve energy efficiency and reduce fuel consumption, based on the optimization of the way (AI) non-manual driving (AI) vehicles are performing. Such as recharge batteries and electric engine will be predicted the popular fuel in order to limit fuel consumption to future (AI) non-manual driving vehicles. They can reduce air pollution, consume less fuel for (AI) non-manual driving vehicles.

(3) Improving comfort by anticipating (AI) non- manual driving vehicle drivers.

Finally, another application for intelligent vehicle is the improvement of driving comfort. Car industry is very competitive market. Many potentials (AI) intelligent vehicle customers need to enjoy to sit more comfortable intelligent vehicles, who will be attracted by (AI) comfortable systems improving when driving, so part of the research in intelligent systems from cars focuses on how to improve the driving experience, i.e. make it easier and more enjoyable, more comfortable to compare to traditional manual driving vehicles.

As an example, lane keeping assistant systems are technologies that actively keep the vehicle in the lane in highways of the driven drifts out of it. Automatic speed regulation keeps the car at a certain speed without requiring to touch the gas pedal. This can be really interesting for, e.g. (AI) non-manual driving truck drivers that spend a lot of time on highways. But these technologies have a limitation in the case of automatic speed regulation, this technology can not copy of a vehicle ahead drives slower than the desired speed, or if another vehicle cuts into the lane.

This case requires the driver to have a constant focus on the road. In order to achieve more comfort, it is better of the system can adapt to changes in its dynamic environment: let the (AI) intelligent vehicle adapt to the speed of the man-manual vehicle, or autonomously change lane when requires. Again, this requires knowledge about the environment, detection capabilities, reasoning and action planning. Intelligent systems can be used in order to create more attractive and more comfortable and more safe, less energy consumption and less fuel expenditure by intelligent vehicles.

On conclusion, hence, it seems that non-manual driving vehicle will be popular to be accepted to replace manual driving vehicle , even non-manual driving public transport will be also popular to replace manual driving public transport, e.g. ferry, rail, underground tran , tram, bus etc. different public transport tools in future global societies. So, creative technology will be applied to future non-manual driving vehicle public transport and private transport tool.

TWO

CREATIVE TECHNOLOGY DEVELOPS TO SPACE TOURISM ENTERTAINMENT

● Psychology and economic environment changing both factors influence whole space tourism market leisure desire

How can psychology method predict space tourism leisure desire? I believe that it has relationship between the space tourism planner and the economic environment as well as his/her psychology as below:
Firstly, on the economic environment influence hand, it includs these both economic situations, either in the good economic environment, many people can earn high income and employers can supply many job number to provide to many people to work, then it will influence the space travelling planner has more space travelling desire. Otherwise, or in the bad economic, less people can earn high income and employers can not supply many job number to provide to many people to work, it will influence the space travelling planner has less space travelling desire.
Secondly, on these both the space travelling planner individual psychology influence hand, the space travelling planner will have these both aspects of

individual psychological influence, it includes these both either positive or negative psychological influence aspectsas below:

On the positive psychological influence aspect, if the space travelling planner has confidence to the space travelling leisure company can provide safe, comfortable, good quality of one space travelling trip arrangement, good taste food arrangement, reasonable space ticket price and every reasonable space trip for space hotel living arrangement and space garden and space farming land visiting journey arrangement, even, space swimming pool and space sport centre and space cinema leisure arrangement to let whom to stay on the planet at least one day trip, it means not one short time space trip, e.g. the spacecraft only flies about half hour or one half. It can not fly to the planet to arrive its space station destination to stay to let the space travelling planner to live at the space hotel at least one night. Then the space travelling planner will have more desire to choose to catch the space tourism leisure company's spacecraft to travel to space.

Otherwise, on the negative psychological influence aspect, if the space travelling planner lacks confidence to the space travelling leisure company can provide safe, comfortable, good quality of one space travelling trip arrangement, good taste food arrangement, reasonable space ticket price and every reasonable space trip for space hotel living arrangement and space garden and space farming land visiting journey arrangement, even, space swimming pool and space sport centre and space cinema leisure arrangement to let whom to stay on the planet at least one day trip, it means not one short time space trip, e.g. the spacecraft only flies about half hour or one half. It can not fly to the planet to arrive its space station destination to stay to let the space travelling planner to live at the space hotel at least one night. Then the space travelling planner will have less desire to choose to catch the space tourism leisure company's spacecraft to travel to space.

Hence, it seems economic environment changing factor and the space travelling planner's confidence factor to the space tourism leisure providers will influence the whole space travelling market whose space travelling consumer's space travelling leisure consumption desire to be more or less. So, any one space tourism provider can not neglect these both factors how to influence whose customer consumption desire.

● space tourism strategy

Future any space tourism leisure business needs have good business plan to outline the space tourism leisure business in these aspects , such as:

different space tourism destinations of every space tourism journey, technical , financial and regulatory factors for growing space tourism leisure consumption into any one kind of unique artificial intelligent space tourism journey for identified passenger target group.

All how to design one space tourism business development plan to attempt to predict whether what trends will influence how every different kinds of identified space tourism journey in order to achieve passenger number growing aim as well as how to achieve one attractive space tourism leisure to satisfy future space tourism passenger individual space travel needs more easily.

I shall indicate what aspects to future every space tourism traveler who will consider in order to reduce the space tourism traveler personal worry to catch any pace boats to leave our Earth to fly to other planets to travel.

I recommend that any space tourism leisure organizations need to concern these aspects in their space tourism leisure business plan as below:

(1) safe space tourism journey

On first aspect concerns safe space tourism journey plan to let all space tourism travelers will considerate safe issue. They must ensure space boats that is safe to catch them to fly to planets in their space journeys. So, any space tourism leisure business will utilize previous flight rated and proven technologies to form the basis for manufacturing spacecraft vehicles, and will incorporate the latest modern avionics and flight systems for ensuring safety, reliability and economical operation in order to reduce any space tourism traveler personal worry to catch any spacecraft.

So, the space tourism safe journey plan is one very important factor to influence space tourism consumer number for them if any one of space tourism leisure business hoped they can grow the space tourism consumer number for long term. For example, the space boat flight hardware must often be maintained at the space station. It is needed to be considered by space boat experts as risky, extremely expensive and potentially sensitive. To aims to ensure spacecraft will offer an economical and safe alternative for any satellite manufacturers and other space tourism entertainment organizations have a desire or requirement for space tourism flight.

(2) reduction cost expense plan

On second aspect concerns reduction cost expense plan, any space tourism entertainment organizations need have the experience and capacity for safely launching a fully loaded , including space tourism passengers and passenger individual cargo for every spacecraft tourism journey. As a result of outsourcing the launch role to a major contractor, the space tourism pilot can concentrate on space boat crews flight training, planning space tourism passenger cargo capacity and preparing space flight manifests , and will as a result, avoid the expense of maintaining a launch operation on a daily basis. In addition, by outsourcing the spacecraft manufacturing, it can avoid spending millions of dollar on facilities and equipment infrastructure and engineering manufacturing expertise.

(3) achieve any space tourism mission plan
On third aspect concerns how to achieve any space tourism mission. Every space tourism mission must be ensure that reliable service is provided to satisfy every space tourism passenger personal space traveler needs and let them to enjoy in their whole space tourism journey, let them to catch a big aircraft in comfortable environment of technologically sophisticated space boat, reasonable and competitive every time space tourism flight ticket price plan is developed and properly revised every time space tourism ticket price when performing their assigned every different space tourism journey mission.
Hence, the space tourism leisure company will provide one careful selected space tourism destination , e.g. Mar planet space tourism journey, Moon planet space tourism journey or no any space destination journey, it means that the space craft only needs to fly one circle around between Earth and Moon space journey etc. that are capable of meeting the requirements of travelling into Earth orbit. So, any space tourism journey must emphasize affordability, reliability, safety, customer service and responsiveness in responding to every client's space tourism journey requirements. Hence, any one of space tourism journey must have clear space journey mission and objective to satisfy any space traveler client target needs.

● Methods to raise space
traveler number

Future space tourism will be one kind of new travel leisure market for any new space travel leisure companies to enter this undiscovered market in the beginning. However, how to predict future 10 to 20 years , even more space traveler number that is one important issue to any new space tourism leisure companies.
I think that space tourism leisure companies need to define what kinds of space travel leisure service to be provided to space travelling passengers,

however, what age group of space passengers who will be their space travelling target client. For example, their space travel leisure must provide any flight operation that takes one or more passengers beyond the altitude of 100 km and thus into space to let space travelling passengers who have fun, exciting space travelling feeling.

Anyway, for any kind of space tourism (leisure space travel) journey, space tourism leisure company needs anyone to be bring customer satisfaction, it is a plan or predictive methods to measure how to let every space travelling passenger to feel comfortable when they are catching the spacecraft (space flying product) and they can have enjoyable and fun or exciting feeling when they have need providing any space tourism journey, services meet or surpass customer expectations.

Thus, any space tourism leisure company needs to evaluate the degree of every time space tourism journey's customer satisfaction and customer satisfaction is also always evaluated in relationship to the every time ticket price of the space tourism journey. So, the space tourism leisure company will predict the next time of what the space tourism journey of passenger number is more accurate, after it has evaluated what degree of every time space tourism journey's customer satisfaction is. It aims to gather their opinions to find which aspects that they need to revise, e.g. choosing where will be the next time space tourism journey destination, how to improve spacecraft staff's service attitude and performance to serve to their space tourism passengers when they are catching the spacecraft, to evaluate whether the spacecraft can provide comfortable and safe environment to let them to catch in order to let the next time space travelling passengers can feel satisfactory and enjoyable when they are catching the space tourism leisure's spacecraft to fly to anywhere in space.

In general, the expectation of factors space passengers include the following customer value elements, such as below:

- viewing space and the Earth.
- experiencing weightlessness and being able to float freely in zero gravity.
- experiencing pre-flight astronaut training and related sensations.
- communicating from space to significant others.
- being able to discuss the adventure in an informed way.
- having astronaut like documentation and memorabilia.

These objectives need to be combined with, sometimes conflicting constraints, such as guaranteed safe return, limited training time, reasonable comfort, and minimum medical restrictions. All these above issues which will be every space travelling passenger considerate matters before they choose the space tourism leisure company to catch its spacecraft to fly to space. So, all these factors will influence the next time space passenger number. Any space tourism leisure company can not neglect how to solve these all matters before they decide when their next time space tourism journey to be achieved.

Consequently, if the space craft tourism leisure company could revise what aspects of its last space tourism journey to find what are its wrong or weakness or unattractive challenges to cause any one space travelling passenger who feels unsatisfactory. Then, it can have more effort to concentrate on improving its next space tourism journey to raise its space tourism service performance level , e.g. people, food, leisure etc. service aspects and its space tourism product quality level, e.g. proving comfortable spacecraft facilities to let space travelling passengers to catch in whole spacecraft tourism journey. Then, it will have more confidence to achieve the raising space travelling passenger number.

● What is the prediction space travelling passenger desire method ?

The prediction space travelling passenger individual desire method can be one survey investigation method. When every time spacecraft finishes space tourism journey mission, after all space tourism passengers catch the spacecraft to arrive earth from space. When they arrive earth space station destination, then the space tourism leisure company can arrange survey investigation staffs to enquire their feeling for this time space tourism journey immediately.

The survey content can include as below:

Do you feel satisfactory or unsatisfactory to which aspects of this time space tourism journey?

(1) On service aspect questions include as below:

(a) Do you feel space food taste is good?

(b) Do you enjoy this time space tourism journey arrangement?

(c) Do you feel satisfactory to space staff
service performance?

(d) If you have unsatisfactory feeling for any one of above questions, which aspect issue cause you feel unsatisfactory to explain to let us to know in order to us to revise our service performance.

(2) On product aspect questions include as below:

(a) Do you feel comfortable when you are catching our spacecraft in whole space tourism journey?

(b) If you feel comfortable , may you explain the reasons what aspects of our spacecraft has weakness to cause you feel uncomfortable?
(c) Do you feel safe when you are catching our spacecraft in whole space tourism journey?
(d) If you feel unsafe, may you explain the reasons what aspects of our spacecraft has weakness to cause you feel unsafe?

Finally, we thank your ideas to be given to let us know how to improve our every time future space tourism journey in order to find what challenge cause our service performance and product quality which can not satisfy your needs. So, we shall improve to avoid future challenges continue occur. Our mission is achievement of 100% satisfactory level to our every space travelling passenger individual feeling. Also, we hope that you can choose our space tourism leisure service again, when you have another time space tourism leisure desire need. However, we shall revise to improve our service performance and product quality to be better, after collecting your ideas from this time survey investigation. I think you spend time to give your ideas from this survey investigation faithfully.

So, survey investigation method will be one important idea gathering tool to help any space tourism leisure company to revise the weaknesses to raise or improve future every time space tourism journey service performance and product quality to achieve raising competitive effort in this new space tourism leisure market.

Hence, survey investigation method will be the best idea gathering method to predict how space travelling passenger emotion or desire need will change in order to achieve the objective of raising every time space tourism journey future space travelling passenger number more easily for every space tourism leisure company.

- The prediction of price factor influences space traveler number

The space tourism leisure organizations indicate the total cost of a trip into space is rapidly coming down from the initial price level of about US$600,000, it is obvious that the space travelling customer base is going to be rather small. Typical customers tend to belong to the top 1% income bracket. They also indicate that the price comes down , it is expected that new space travelling customer groups will enter the space tourism leisure market.

Typical new customers include people in other brackets with one-of-a kind incomes, such as inheritance or business sold. There are indications that those types of customers are becoming interested in spending on an once-in-a lifetime space experience. Therefore, the growth of the space tourism market is highly sensitive to customer satisfaction and how it is communicated through various media.

This will establish the status factors of space tourism and corresponding brand reputation service providers. They also suggest that any operator monitors space travelling customer satisfaction closely, as it will help developing increasingly accurate estimates of how the space tourism leisure market will develop.

Hence, it seems that every time space tourism journey price variable factor will influence the time space tourism of customer individual leisure desire and the space tourism passenger number. For example, the minimum price goal foe a variable space tourism business is currently estimate to be below US$3000-4000/kg for a round -trip depending on variable configuration and operation size. At this price, they estimate that somewhat over 1 % of the high income earners are potential customers.

However, for significant volume growth the longer term goal should be below US$2000/kg for a typical passenger, baggage and supplies. The lower price will probably open space tourism to a broader population, expanding the customer base and altering expectations. beyond this point space tourism will become into a travelling competitive leisure commodity, price competition will ensure and service providers need to rethink their space tourism marketing and branding and price strategies.

I shall also recommend how to attract the potential customers successfully. First, space operators need to pay special attention to the right level of customer services. Second, various preparatory customer operations cost, such as a travel to the launch site, space tourism destination accommodation, pre-flight training, medical check-ups and equipment my add up to between 10 to 15 % of the actual space travel cost. Thurs, solving the right balance between services offered and cost of client operation in order to earn the largest intangible benefits, such as loyalty, confidence, leisure enjoyment, comfortable space travelling journey as well as tangible benefits, such as profit, spacecraft manufacturing facilities, space stations, space hotels , space swimming pools, space gardens, space cinema etc. which are built to similar to earth building facilities to satisfy space travelers' needs.

The influential factors persuade travelers choose space tourism

Nowadays, our earth is no longer an adventurous enough place for some experienced tourists. Space tourism will be a new sector of adventure tourism, which is in the near future will be fast becoming a new tourism leisure opportunity for experiencing the unknown. Of one day, space tourism is able to reach the mass tourism phase, due to improved safety and decreased operation costs, a future space tourist will possibly only need minimal training to cope with the zero cost.

Space tourism is quite well established with visits to space attraction and launch sites, and it is a wealthy trips to the international space station for any space tourism travelers. However, if any space tourism leisure companies can attempt to find what the most influential factors are to persuade travelers feel attraction more than travelling in our earth.

It aims to let travelers to choose space travelling more than earth travelling when they feel travelling leisure need. I shall indicate what will be the most important influential factors to persuade travelers to choose space tourism more than earth tourism as below:

Firstly, I shall argue that the majority of different new space tourism journey destinations will be needed to find to satisfy different aged space travelers and different income space tourism consumers' needs. For example, the rich people have effort to consume longer time and reach any space tourism destinations where are far away from our earth of their every space tourism journey.

Otherwise, the middle income people will choose shorter space tourism journey distance from our earth and short time space tourism journey. Also, younger space tourism clients can accept more longer journey time, exciting fast speed spacecraft flying journey. Otherwise, old space tourism clients can only accept comfortable and shorter time safe space journey. So, it seems that safety, comfortable feeling, shorter time space tourism journey won't be one important influential factor to excite any young people who choose to consume space tourism leisure. Otherwise, safety, comfortable feeling, shorter time space tourism journey will be one important influential factor to excite any old people who choose to consume space tourism leisure.

Secondly, the another most important influential factor to excite space travelers to choose space tourism , it concerns whether the space travelers will feel what tourists benefits can be earned from a substantial variety of destinations choice. In general, space tourism with those of aviation, space travelers will hope space tourism will be travelling distances by air in a very short time, safely and comfortably, to bring them to arrive any space planet destinations when spacecraft reaches any space stations to stay in any space destinations.

Hence, space destination factor will bring important influential choice to any space destination journeys. As a result of the space technological tourism boom, the number of potential different space destination, choice attractions have grown with far fewer places on earth to which human do have access yet. However, the ultimate different space destinations to which many of us dream is not on earth, but as least 100 km above us, anywhere in space any planets.

If the space tourism leisure company can provide different space tourism destination choices to young or old age both space traveler target consumer groups. They will feel a real holiday when they will be able to enjoy a great image of the earth from planets. It might mean that every space tourism journey can provide different space tourism destination to let space travelers have another new travelling destinations where are far from our earth anywhere.

Hence, the different space tourism destinations will give them an unforgettable adventure. Think of how it would be to be able to check in at a " billion strategy" luxury hotel in space one planet, it means that the space planet destination can provide one luxury hotel to let space travelers to live one night or more in the space planet destination, how it would be to schedule the space traveler‘ vacation at one of the space tourism leisure company luxury resorts on the Moon or Mars.

This images seem from science fiction movies, but one should not forget that 100 years ago, the Wright brothers, aviation pioneers inventors and builders of the air plane, would not have imagined how, every day it is possible that future spacecraft can fly to any planets to let human have chance to stay in the space hotel one night or more.

Consequently, space destination choice and space tourism journey service performance, aviation safety, ticket price and leisure satisfactory feeling which will be important influential factors to attract future space travelers to choose space tourism leisure to replace earth tourism leisure in future one day.

● Raising space tourism leisure
consumption strategies

Although, space tourism industry is a real enjoyment and exciting travelling leisure to human. It is possible that human will choose to consume space tourism leisure to replace earth tourism leisure, if human felt that earth tourism leisure is not attractive to them to consume to go to anywhere to travel in their leisure time.

But, I believe that space tourism industry has still many factors to influence human to choose to consume space tourism leisure, even they will consider space tourism leisure consumption I is only one time space tourism in their life time. Hence, space tourism companies ought achieve this aim to persuade or attract everyone prefer to spend space tourism leisure at least one time in their life, then it can represent success. However, I think to achieve this aim, it has these challenges to influence their success, even they believe space tourism leisure business is one potential attractive travel entertainment business. These challenges include such as: expensive space tourism ticket price issue, catching spacecraft safe issue, space traveler personal body health issue, age issue, family and friend relationship influence issue, working time and holiday time arrangement issue, the space trip arrangement issue, weather issue etc. different challenges, which will have possible to influence every space tourism planner either who decide change to cancel the time space tourism plan, or forgive to choose space tourism leisure in their life forever.

Hence, how to raise space tourism leisure consumption desire will be one considerable matter for any space tourism leisure businessmen. I shall indicate my personal three aspect of strategical opinions to let them to know how to raise every space tourism planner individual space tourism leisure consumption desire to avoid every time space tourism passenger number will have decrease failure chance as below:

● (1) Strategic opinion

On the first aspect of strategic opinion, I feel that the space education tutor can teach new space knowledge to let every space traveler to learn any new space and earth knowledge during he/she is catching on the spacecraft in personal contact learning experience environment which can raise space tourism consumption desire. The reason is because the space tourism leisure traveler can raise extra space and earth learning knowledge when they can catch the spacecraft to fly and contact the space environment to learn and feel what the differences are between space and earth by himself or herself. Hence, it is very attractive to the space traveler student target group and I believe that their parents will encourage their sons or daughters to participate the time of space trip and they are more preferable to help them to buy the time space trip ticket, due to their sons and daughters can learn any space knowledge when they are studying. Moreover, every space traveler will feel surprise to learn any new space and earth knowledge from the space tutor's teaching, due to he/she is unknown that this space travel trip includes learning space and earth knowledge.

I suggest that the space tourism leisure businessmen can give learning opportunity to every travel trip space travelers to feel that this space actual environment can bring what disadvantages or advantages to influence our earth when they are catching aircraft to fly to space to travel in every space trip. The space and earth learning knowledge can include these two aspects of space learning knowledge and experience below:

On the teaching of space environment learning knowledge hand, the topics can include as below:

Firstly the space learning topic can concern how space environment influences water and hydrated minerals change , they can learn what our drinking water function how is applied to space environment. For example, in the space environment, they can learn and attempt to feel that how water can be used in protecting astronauts against harmful radiation from the sun and cosmic rays by cloaking spacecraft with a thin layer of water in the actual space environment as well as the space travelers can also feel water is same as fuel when they are catching the spacecraft, they can feel the water is heavy to transport into space when they are catching the spacecraft to fly to space during their whole space tourism journey.

Moreover, when their spacecraft reaches anyone of planets and it stays on the planet's space station, e.g. Moon space station. They can learn how to attempt to contact the hydrated minerals to learn and feel what they contained in some asteroids may be possible sources of water and fuel in the actual space environment. When they are walking in actual space environment, such as Moon planet, they can contact or touch this hydrated minerals to learn how water molecules can be extracted and separated chemically to produce hydrogen fuel knowledge in the actual space environment. This is one exciting space learning experience to the space travelling student passengers.

Secondly the space learning topic can concern how human fights space threats , even when their whole space leisure journey, the space science teacher can let the space trip student passengers to feel that they are learning new space knowledge between the space science teacher and whose space trip student passengers. Such as how to protect our earth knowledge: Teaching them to know when will be threats to our earth from space. The space science teacher can explain how this space threating environment influences our life safety and let them to feel that a mass extinction can be triggered if an asteroid 10 kilometers across hit the earth. Even being the apex species in the food chain did not space carnivorous dinosaurs from such disaster, who knows if this terrifying scene won't happen before our eyes? So, the space travelers can image and feel how the space threating environment can influence their life safety in the actual space environment as well as the space science teacher can let whose space travelers to feel and image the actual earth disaster will possible happen suddenly to let they feel afraid in the actual space environment. Also the space science teacher can teach how our earth can fright the space stones attack to let the space traveler to know, when an impactor targets an asteroid for a controlled well-times wallop. The collision will change the asteroid's momentum, deflecting it from its original orbital path which intersects with that of the earth. So, at the moment, the space travelers can image they are a larger spacecraft near an asteroid which can also change the path. Given enough time, the gravitational pull from the spacecraft will be able to steer the asteroid away from the earth. So, every space traveler will feel that they are catching the spacecraft in the safe space environment to avoid the Earth disaster from space sudden unpredictable attack.

It is more fun real space tourism knowledge learning feel to let every space traveler has chance to learn any new space science knowledge when he/she is catching the spacecraft to fly to space to travel. Hence, one successful space trip ought include trip and learning experience both contents in order to raise every the space tourism planner individual space trip consumption desire.

● (2) Strategic opinion

On the second aspect of strategic opinion, space tourism leisure companies need to let planning travelers feel that anyone of space tourism leisure is very different to general tourism leisure. In general, tourism leisure is visiting at least one night for leisure and holiday, business or other tourism purposes in Earth only. Otherwise, space tourism leisure is other kind of an unique trip leisure or entertainment method, e.g. the space traveler can catch the spacecraft to visit any planets to stay to live at the planet's space hotel at least one night, e.g. Future potential populated Moon or Mars space hotel space trip. Moreover, the space travel companies ought give chance to let them to feel what weightless feeling is in weightlessness environment when they are walking on Moon or other planets in possible. Even, they can attempt to build these entertainment facilities, instead of space hotels, such as space swimming pools, space gardens, space cinema etc. building facilities. It aims to let them to feel what the differences between Earth and space life when they are walking on the Moon, when they are swimming on the space pools, when they are living in space hotels, when they are watching movies in space cinemas, when they are seeing flowers and different species of planets and fruits. e.g. oranges, apples, bananas, and vegetable and potatoes and tomatoes in space gardens. It is very exciting and fun space trip life experience between one days to seven days. So, they believe that they must not feel these space life experience if they do not choose to participate this time space trip planning journey by the space trip company preparation.

Also, due to that the space tourism passengers need to the pre-flight checks and training before they ensure to qualify to permit to participate the space trip. So space travel companies need to concern how to take care their health check and training matter considerately. It aims to let every space traveler will feel a market segment with fitness and extreme experiences as well as he/she will become popular with a market segment passenger to the space tourism leisure company, although he/she must not guarantee to pass the space training and/or pre-flight health checks to permit to participate the space trip. However, he/she can believe that he/she is one worth space travelling passenger to the space tourism leisure company, even this time pre-flight health check or/and the short time space trip training requirements are failure. However, the space tourism leisure company must need to let all pre-flight health check and space trip training passengers to feel that it is only one space tourism which can give them and let customers view the space travel is as the ultimate showcase for health, even though a majority of the population can pass the pre-flight medical and other tests in order to raise their confidence and safety to catch the spacecraft to fly to space to travel when they are confirmed to pass these tests to permit to catch the spacecraft later.

In general, the expectations of future space passengers include the following customer value elements, such as below:

- Viewing space and the Earth.
- Experiencing weightlessness and experiencing pre-flight astronaut training and related sensations.
- Communicating from space to significant others.

- Being able to discuss the adventure in an informed way.
- Having astronaut-like documentation and memorabilia.
- Enjoying one exciting and fun space trip.

However, instead of considering these objectives need to be combined with, sometimes conflicting , constraints such as guaranteed safe, return , limited training time, reasonable comfort, and minimum medical restrictions. So, space tourism companies need to reduce every space traveler individual worries before they decide to make the time of space tourism journey. Then, it can increase their confidence to raise their space tourism consumption desire more successfully.

Consequently, instead of these consideration, a space travel operator must pay attention to the total customer experience over the entire customer process, starting from how the service is presented, proposed and sold. The service package must include training, instructions, travel to the launch site and various post. Travel activities to generate maximum customer satisfaction and brand building opportunity.

(3) Strategic opinion

On the final aspect of strategic opinion, I think any space tourism companies space tourism companies need to consider every time space tourism ticket price and space tourism trip issues. It is important factor to influence every space traveler individual consumption desire. Due to space trip ticket price must be more expensive to compare common Earth trip travelling ticket price, so this kind of tourism leisure market target customer will be the rich and high income customer group.

On the space trip ticket challenge issue, despite that fact the total cost of a trip into space is rapidly coming down from the initial price level of about US$60,000, it is obvious that the customer base is going to be rather small and the client target customer is only high income or rich consumer group. Typical customers tend to belong to the top of the top 1% income bracket. So, ensures that space traveler number must be less than common Earth traveler number.

Also, such as the space trip ticket price, it is expected that new middle rich level or middle high level income customer target group will enter the space trip leisure market, when every space trip ticket price falls down about 1% Typical new customers include people in other income brackets with one-of-a-kind incomes, such as inheritance or business sold space traveler target group. These people will be space travel new client group, when its every space trip ticket price can be reduced to close 1 to 2 % nearly. If any space tourism leisure companies expect to attract new rich and/or high income target customer group to choose any one kind of space trip journey planning to consume.

These are indications that these types of customers are becoming interested in spending on an once-in-a-lifetime space experience. Therefore, the growth of the space tourism market is highly sensitive to customer satisfaction and how it is communicated through the various media. This will establish the status –factor of space tourism, and corresponding brand reputation of service providers. The minimum price goal for a variable space tourism business is currently estimate to be below US$3-4000/kg for a round-trip depending on vehicle configuration. So, space travel leisure companies need to concern every round space trip cost, it can depend on the space vehicle number and weight issue to influence every space trip ticket price variable to achieve how much it can earn.

On space journey design factor aspect, it includes these different facilities aspects how to design, because future space travelling consumers will concern whether the space travel company can provide special entertainment to satisfy their needs. The facilities include as below:

How to design space hotels to let them to live in comfortable space environment and eat the best taste and fresh food quality when the cookers need to cook in the space hotel in the space environment? How to design space swimming pools to let them to swim in safe space environment? How to design space sport centers to let them to run more easily in one space sport warm and safe environment? How to design one space garden to let them to see different species of Earth flowers, or plants? How to design one space farming land to let them to see different species of Earth fruits, vegetables, tomatoes, potatoes etc. fresh foods growth in warm and safe space farming land environment? How to design one space cinema to let them to watch movies in one safe and warm space cinema environment? All these facilities will be any one of future space trip's' important and attractive space trip leisure facilities to influence every space traveler to choose to buy the space tourism leisure company's space trip leisure service. Instead of these space building entertainment facilities, they also need to concern how the space vehicle entertainment tools are provided the entertainment service to satisfy their needs. When the space travelers can sit on the space vehicles to move on any planets' lands, such as Moon. A number of space vehicle options exist in the market, mainly differing based on the seat capacity as well as the in-flight experience level offered. The

typical space vehicle solution is a small, relatively light weight spacecraft taking between 2 to 10 passengers. The number of passengers depends on the service level, amenities and extra offered. The trip typically lasts about 10 hours and of which about 4 hours are spent in space. The main attraction is the weightless time after in space. The main attraction is the weightless time after re-entry has started. It is a rather low-G technology and therefore the medical requirements for participants are nor very high.
Consequently, the space vehicles, space leisure building facilities, the space trip reasonable price ticket level, every safe space trip journey arrangement, clean and fresh and good taste space food arrangement, space traveler individual real learning experience etc. these factors will be the main influential factors to raise the space tourism leisure company's competitive effort and the space traveler consumer individual consumption desire to the space tourism leisure company in the future.

Research how to raise space traveler individual leisure desire
The first factor may raise space traveler leisure desire is that space rocket needs have safe and clean inside environment. Future, space tourism may be another kind of possible popular tourism leisure activity, because since COVID 19 disease occurs, it may influence many travellers feel afraid to catch air planes in the high risk closed window airt plane inside environment, when many different countries people must need to body contact or air contact. Hence, it is possible that our tourism leisure will not be popular in our earth, if COVID 19 disease , even other unknown air disease may occur to cause global travellers feel fear to catch air planes to avoid life danger in the one hour, even more than 12 hours sitting air plane flying time. Otherwise, because any space tourism rockets are small and it only allows one to four space travellers and space rocket pilot to sit in the space rockets as well as any space journey is only spent 15 minutes to 30 minutes for view moon space journey or more than one day visiting moon space journey or up to one week visiting space station journey in the space rockets. So, owning COVID 19 disease travellers can permit to sit in the space rocket, the chance is low to get COVID 19 disease when the space rocket has less passengers are sitting in the space rocket. It means that when many space travellers feel any space rocket is safe and clean in the inside none window space rocket environment, this safe and clean space rocket inside environment feeling, it cam emcourage many future space passengers begin to choose to catch any one space rocket to leave

earth to fly to space to travel in possible. Hence, when space rocket can be invented to bring equipment safe and comfortable feeling and guarantees none of any one owning COVID 19 disease patient can sit in the space rocket as well as all of they can catch this space rocket to earth and come back to earh safely . It means that any one space tourism leisure service provider can guarantee without any sudden accidents occurrence in space. The another most important factor may be every space tourism can charge reasonable ticket price to any one space trip to let any one space tourism leisure consumer to feel. All of these factors may influence the space tourism leisure consumers number increases to the space tourism leisure service provider.

Instead of above space trip ticket price and space rocket's safe and clean inside environment both aspects. I shall indicate how to raise space traveller individual leisure desire methods as below:

Firstly. attractive space trip is another important factor to influence any one space tourism leisure consumers to choose the space tourism leisure service provider's any one space tourism leisure activities. Because of the space trip's time is short, e.g. 15 minutes to 30 minutes leaving earth to stay on space for viewing our earth leisure , within this 15 to 30 minutes short space trip time, the space tourism leisure service provider needs to seek the different kinds of attractive space destinations to let every space tourism leisure passengers to feel enjoyable in space different locations when they see our earth from their rocket's staying different locations in space. Because different space locations, they can influence different viewing feeling of our earth from the rocket. So, choosing the suitable different space locations to stay in order to view earth , this short time space staying locations trip is very important to influence every passenger whose viewing earth feeling. Otherwise, if the space trip's time is longer, e.g. more than 5 hours sitting to visit moon space trip. The space trip must need to be arrangement have more attractive feeling to let they do not feel boring when they need to sit more than 5 hours in the space rocket to arrive the moon. So, their sitting space rocket times are needed to be feel leisure times for every passenger. They can not feel bore or non space trip arrangement leisure feeliing in the whole space trip. So, when the space trip is longer time, its space leisure activities arrangement ought need have enough different leisure activities to provide in order to avoid any one space passenger feels bore, e.g. spending one week to arrive space station space trip, the rocket must need have sport equipment or cinema provide to let they do not feel bore when they need to sit about one week time to go to the space station as well as spend another one week time to come back earth from the space station. Otherwise, short time space trip , e.g. only 15 minutes viewing earth space trip , it doe not any space trip leisure activities, because this space viewing earth trip aims to let passengers can feel comfortable and enjoyable to view our earth when they are sitting in the rocket in this 15 minutes. However, this rocket needs to fly to about 4 to 5 different space locations to let passengers to feel different viewing earth visable feeling. If the rocket is only staying in one same space location to let them to view our earth. They must feel bore to view the same visable feeling of our earth in this 15 minutes space viewing earh trip. So, space locations choice factor is very important for short time viewing earth space trip. Consequently, they may feel their this short time viewing earth space trip ticket price is unreasonable.

Hence, any one space trip lesiure activities and space staying locations choice arrangement, as well as its space trip time as well as its evaluated ticket price, they must have close relationship to influence any one space trip traveller individual leisure satisfactory feeling in this space tourism leisure development industry. Hence, if any one space tourism leisure service provider hopes that they can have many space travellers to choose their any one long or short time space trip lesisure activities. They must need to consider how to design space trip in order to attract their leisure choices , how to evaluate every space trip's time and space trip ticket price in order to achieve how to attract many future space traveller individual preference space trip choice among potentient different space trip lesiure providers. Because when future space tourism leisure is popular to accept. In this space tourism market, it will have many potential space tourism leisure service providers attempt to anticipate and implement any kinds of space trip leisure activities arrangement, it seems future space trip may influence our traditional tourism industry from earth travel changes to space travel lesisure direction development.

In earth tourism industry, every travel leisure service provider needs to design different kinds of travelling destinations in order to attract different countries travellers choose to play themselves designing trips packages, if their trips arrangements are attrractive, it will influence thr traveller chooses this travel agent's trip to visit the another country service. Hence, earth trip's travelling packages are very important to influence every traveller individual travel agent choice. It seems that space tourism trip arrangement may be another main factor to influence space passengers number increases or decreases to any one space leisure service provider. The question concerns how to design attract space trip to any one space traveller choice preference. I shall explain as below:

IN fact, space tourism's general price is expensive, so the wealthiest people ought be any ones space tourism leisure provider's main customer target. But advances in rocket and capsule design also also expected to lower the price to the point that people of more modest fortunes are able to afford a ticket. What space tourists can expect? What exactly is on store for space tourists? The excitment of a rocket ride and a chance to experience weightlessness for starters. And bragging rights are hard to beat. But some the biggest benefit of point into space is getting a dramatic new outlook on life when any one space traveller can try to catch rocket to leave our earth home.

For Virgin Galactic plans to offer suborbital space trips, with customers being treated to feel of weightlessness feeling. He says more than 600 customers have signed contracts to already to pay a ticket price US$250,000 for one time space trip. Another space tourism leisure provider, e.g. Blue Origin, Amazon CEO Jeff Bezos , thees space tourism leisure providers had begun to plan future space adventures leisures to satisfy future any one space traveller individual space trip leisure need. In general, average a ticket price is between US$75,000 to US$300,000. SO, wealth people must be space travel consumer targert customer.

Future space trip may include: Flying to space stations, it may be a big vacation will be able to buy a rocket ride into orbit, future NASA is not transfoming into a space travel agency, private companies will have to pay it about US$35,000 a night per passenger to sleep in the space station's beds and use its amenities, including air, water , the internet and toilet. Hence, flying to space station may be one attract one to two weeks attractive space trip.

Another kind of space trips, such as a variety of options for private of spaceflight have started to emergy. Virgin Galactic, founded by the entrepreneur Richard Branson, and Blue Origin from Jeffrey P. Benos of Amazon, both plan to carry passengers on short suborbital flights, space X also announced that Yusaka Maezawa, a Japanese, clothing company founder, would pay for a trip around the moon on a spacecraft it is building. Hence, short time space trip, e.g. flying to moon or leaving earth short suborbital flights, staying on space to viewing our earth , even long time space trip, e.g. flying to space station,spends more than one to two weeks. They may be future any one wealth space traveller's space trip leisure choice.

Some companies already conduct modest experiments on the space station, such as Merall Research laboratories, which has grown crystals of antibodies, and mode in space, which is testing the manufacture of higher quality optical communications fiber in the weightlessness of orbit. Hence, flying to moon and staying on space to viewing our earth space trips both ticket prices miust be more cheaper to compare long time spce trip, e.g. visiting space stations space trip. Axion space, a houston based company that arranges training and all aspects of the flights, is charging as much as US$55 million for a week long trip to the international space station. Blus Origin , Virgin Galactic had been planning attractive space trip. It focuses on views of earth space leisure activities when its space travelling customers can sit in its space rocket. Virgin Galactic was charging as much as US$250, 000 per seat on its spaceship. HOwever, these both space planes have a waiting list of about 600 passengers.

ON conclusion, in the future, the different kinds of space trips may include: short journey viewing earth, 15 to 30 minutes, visiting to moon, about one days , even it is possible that visiting space station, one to two weeks. Even it is possible that any one space traveller can attempt to live one night in space hotel, if human can build hotels on moon, when human can confirm moon can build hotels , then flying to moon to live one night space trip experience may be implement in possible. I believe that any one space leisure provider can attract many space passengers to choose their viewing earth short time staying space trip, visiting moon, living space hotels, visiting space stations to live one night space leisure trip, So, any one space trip leisure provider can attempt to evaluaate whether implementing "living moon hotel" or " visiting space station" or viewing our earth any one space trip is possible in order to raise future any one potential space traveller individual leisure desire. However, the main considerable points, they need to evaluate whether they ought charge how much space trip ticket or seat price for every passenger. Also, they need build confidence to let every passenger feels their space rocket is safe to sit to leave our earth and come back our earth again absolutely. Hence, all of these are important factors to influence future any one space tourism leisure provider their success, if they expect to develop their space trips businesses in long time.

- Space tourism leisure behavioral economic

consumption model

In space tourism leisure industry, due to every time space trip needs the space travelling planner to plan how much budget to consume expensive spce ticket price. So, it seems that the target customers will be rich or high income level young people or the retirement rich old people target customer group.

So, it brings this question: How to persuade these rich or high income young people or rich retirement old people to prefer to spend spce tourism leisure at least one time in their life?

It is one valuabe research question to every future space tourism leisure provider. I shall indicate the successful factors to analyze how to persuade them to accept this kind of potential space travelling leisure in behavioral economic personal consumption view point, in order to explain the cause and effect relationship between of these factors as below:

(1) Economic environment variable factor

Firstly, it is economic environment variable factor whether it can influence to space tourism leisure consumption changing. As I discuss about economic environment variable issue will influence consumption behavior changing. For space tourism leisure case, it is not now kind of essential consumption leisure product to every one. So , even the rich or high income people who will be influences to seek this kind of leisure to play, it the economic environment is improved, it will influence they have positive attitude and interest to choose this kind of leisure consumption. However, if the economic environment is worse, it will influence they have negative attitude and no interest to choose this kind of leisure consumption, due to space travel is one kind of expensive leisure consumption to every one.

Hence, in this space tourism leisure industry, it does not ensure that the rich or high income people must be persuade to choose this kind of expensive space tourism entertainment in whose holiday or retirement time. They can have the common tourism entertainment to go to different countries to travel many times in our earth. Otherwise, space tourism leisure is more expensive to compare common earth tourism leisure , it means that the rich or high income people only spend one time spacecraft catching to fly to space to travel in their life, it is more difficult to every space traveler like to catch spacecraft to fly to space to travel more than one time, due to he/she had attempted to catch spacecraft to fly to space to travel to own space travel experience, he/she will feel enough satisfactory and enjoyment in common. Hence, it is possible that future many rich or high income people only like to spend one time space tourism leisure, then they won't continue to spend this kind of tourism entertainment again in their life.

Thus, space tourism leisure providers need to arrange any special or attractive space tourism leisure to persuade these high income or rich target clients to consume, when the economic environment will change worse. The Europen space agency (ESA), defines this phenomenon between economic environment variable and space tourism client growth or falling number relationship as: " space tourism is an execution of sub-orbital flight by privately finded and/or privately operated vehicles and the technology development driven by space tourism market."

it seems that space vehicle is one attractive travelling desire tool will be one attractive selling point to influence space tourism leisure consumer individual entertainment choice or attitude to be changed to positive leisure consumption attitude to prefer to play this kind of space tourism activities when economic environment changes to worse. Hence, when economic environment is worse, the economic wore changing factor will influence the space travelling planner individual leisure consumption desire, even it will influence the rich or high income young people or rich retirement people target customer both groups.

As (ESA, 2008) indicated space vehicle will be one kind of attractive leisure tool for spce traveler. So, I suggest that space tourism lesiure journey arrangement needs to include that such as : the space travelers can catch space vehicle to move on Moon or Mars plants land to feel what the different feeling is between during they are catching public transportation tool, such as bus or taxi during the are catching these transportation tools on earth land and during they are catching space vehicle tools on Mars or Moon planet's lands. It is so exicting and fun catching space vehicle tool experience on these both Mars or Moon planets' lands to the young and old age space travelling passengers. Because every space vehicle's speed is not very fast and it will move on Moon or Mars planets slowly. So, any aged pace travelling passengers can attempt to play this kind of space facilities leisure after they catched spacecraft to fly to these both Mars or Moon planet to stay. They can spend half hour or one hour, even more than one hour to catch the space vehicle to go to anywhere on Mars or Moon to travel. It is possible that they can find exciting and undiscovered things on these both planets.

So, catching space vehicle to go to anywhere on either these both planets journey, it will one essential part of space travelling journey during the economic environment is changed to worse. It is extra attractive space travelling leisure journey to attract space tourism consumer individual leisure desire when economic environment is worse.

Hence, from this perspective then space tourism could be understood as a section of the tourism industry mainly based on technological development, progression and its activitity being related specifically to sub orbital flights. So, if future space tourism providers expect whether the global economic environment changing will be better or worse which won't influence space tourism leisure consumption desire to be changed. The space tourism leisure providers need to persuade the space tourism planners feel space tourism would have to be treated like an already exciting part of the tourism industry. It means that space tourism leisure is one kind of tourism leisure choice to replace common earth tourism leisure consumption. When travelers feel space tourism is another tourism leisure to replace which can replace common earth tourism leisure. It will avoid the worse economic environment changing factor to reduce the rich or high income young people or rich retirement old people whose space travelling leisure consumption desire.

Consequently , the question in relation to, in what kinds of space tourism journey message do space travel providers promote behind whether space vehicle journey promotion message which is needed when economic environment will change worse. I shall be asked, as understanding the meaning in which space tourism is being marketed, communicated is seen as a factor , which can either positively contribute to future development of the tourism industry or lead into prolonging or seen stopping the space tourism industry from its progression.

(2) Space tourism leisure journey management factor

Secondly, space tourism leisure jounrey management factor, how to arrange every space tourism leisure journey which will be one important factor to influence space tourism planner individual tourism consumption desire.

In general, it can includes these several forms of space tourism leisure activities in every space lesiure trip arrangement. The following classification of space tourism include: Terrestria spce tourism (i.e. NASA visit centre, space movies, online space experience); Atmospheric space tourism (i.e. : MIG 31 flight, zero G. flights) and astro (orbita) tourism (i.e.: trips to the international space station-beyond earth orbit) (Cater 2010,

Crouch et al. 2009).

Instead of US domestic space tourism market is potential, next country is Japan. First, the study is made by Collins et. al (1994, 1996) in Japan on 3030 research participants, showed that 80% of respondents under the age of 50 were willing to travel to space and out of them 20% were willing to pay year's salary for the space travel experience. Yet, it could be citicized that the Japan people age group of under 50 could be too broad, in general different generations under one groups. nest besides the willingness to go to space, the Japanese study showed respondents motivations for travelling to space, including any fun and exciting attractive space tourism journey, e.g. interest in space walk, catching space vehicle or driving space vehicle on the either Moon or Mars planets, earth view, zeo gravity experience, livin gin space hotels one night or more, watching movies in space cinemas, swimming in space pools, visiting space gardens, running in space sport centers, catching spacecrafts to view earth or Moon or Mars planets.

Hence, it seems attractive space tourism journey can persuade another country's space travelling planners, such as Japanese attempts to satisfy whose space tourism needs. So, different kinds of attractive space trip journey arrangement will be one important factor to influence young and old age travelling consumption desire. It implies that attractive space tourism journey will be one influential factor to encourage other countries tourism consumers attempt to another kind of leaving earth tourism leisure.

So, any space tourism trip destinations and leisure facilities arrangement must need to satisfy space traveler individual leisure needs and every space trip must be more fun, exciting and comfortable and enjoyable feeling to compare general tourism journey in earth. Due to general earth tourism leisure will be space tourism leisure's competitive or replaced leisure product and service. Hence, space trip destinations and leisure facilities choice will be one important factor to influence space travelling planner's consumption desire.

Every space travelling planner will compare general earth travelling leisure's destinations and leisure facilities arrangement whether the space travelling trip arrangement , leisure facilities arrangement and food arrangement, space vehicle or spacecraf leisure comfortable influence issues which will have more satisfactory enjoyable feeling to compare general earth tourism leisure and their spending expenditure to every space trip whether is value or is not value.

Consequently, economic environment changing factor and space trip and leisure facilities arrangement factor which both will influence any space tourism planner individual consumption desire mainly. So, space tourism businessmen ought concern these two aspects of factors how and when will change to adapt any country's potential space traveler's space tourism changing taste and needs in order to follow the new space tourism changing needs easily.

Space tourism market moral ethic risk threats

What are space tourism moral ethic risk during the space businessmen operate this businesses as well as what market threats who will encounter to face difficulties ? I shall give actul cases to explain how and why these challenges will cause to influence any new space tourism businesses development successfully.

(1) Potential accidents aspect

Firstly, space travelers will concern that public reactions to potential accidents aspect during they are catching spacecrafts to travel to space. In fact, it is moral ethic responsibility to any space tourism leisure providers to provide safe, comfortable and non accident occurrence in their whole space trip. Because once time accident will cause any one of space passenger hurt or death. So , it must be any space tourism businessmen responsibilities to concern whether they have enough confidence to ensure none any accident occurrences in every space tourism trip.

Hence, in space tourism industry, government needs have public policy to threaten or prohibit any space tourism leisure providers neglect to often check and ensure any spacecraft machines or equipments are regular opeations, as well as often renew new spacecraft machines when they are old to be used. The policy is a force effort to need them to abide every space tourism leisure safe responsibility to ensure or guarantee any one of spacecraft won't have accident occurrences during it has left earth to fly to space in whole space trip journey from the beginning to the end till to the spacecraft come to earth safely.

Hence, this policy forces any space tourism leisure providers concern to put a monetary value on increased or reduced risk of death, the " value of statistical live", used to characterize when the benefit of safety regulation is worth the cost such regulation improves. So, the country government and the country's space tourism leisure providers both have responsibilities to guarantee all space tourism passengers' life safety. It must not allow any death or hurt occurrences during every space tourism trip.

Even, the country government can have legal action to publish any space tourism leisure providers, when their every space tourism trip has occurred accidents, e.g. fire accident occurrence in spacecarft or spacecrat machines are broken to be damaged and need to be repaired during the space tourism trip. It will threaten to reduce trip accident occurrence, such as this cases. The commercial space ventures may present risk to property as well, such as a fire starting on the ground by launch-related material or problems presented by space debris.

In principle, liability law can provide incentive to deter carelessness that could lead to the destruction of property, although statutory (rather than common law) assignments of liability for commercial launches are somewhat problematic.

Consequently, if the space tourism leisure provider expected to grow space tourism passenger number in long -term time, it must need to ensure none any accidents can occur during any space trip. Otherwise, the space tourism passengers can choose another space tourism leisure provider to replace its spce tourism leisure easily.

(2) Space tourism destinations and space tourism entertainment facilities safe arrangement challenges aspect

Secondly, it is space tourism destinations and space tourism entertainment facilities safe arrangement challenges. Nowadays, commercial space travel is looking more like a real possibility than science fiction. The usual ethical issues related to the safety of the space destination choices and the space tourism entertainment facilities, e.g. space vehicles, space hotels, space swimming pools, space sport centers, space cinemas, space gardens, space farming lands. In this strange space environment and safety concerns are just the beginning as there are othe interesting questions, such as below:

What likely would be a fair process for commercializing or claiming property in any space planets? Such as Moon or mars, when any future space tourism leisure providers who need to build above these any one of space entertainment facilities on these planets to provide to their space travelling customers to play.

How to distribute and manage these any lands ownership to these future space tourism providers fairly and legally?

How likely would a separatist movement be among space settlements to want to be free and independent states?

How to ensure above future space entertainment facilities and space entertainment places are in the safe space environment to be provided to any space travelers to play in any planets, e.g. Moon or Mars etc. planets.

So, concerning how to arrange space entertainment facilities to provide to space tourism clients to play in any safe space environment issue, it will be another concerning question to every space tourism leisure providers. When they decide to choose anywhere to the space hotels, space swimming pools, space gardens, space cinemas or space farming lands or space sport centers. These space buildings will need to be built in the safe, on stable stone lands environment and none any natural distaster, such as large wind or space underground water etc. unpredictable space natural distasterr attack to these space buildings suddenly. Because it has responsibility to any space tourism leisure providers to guarantee any one of these space buildings are safe to be built in the planet's safe land environment. It aims to achieve none any accident occurrences during their space tourism clients are staying to enter these any one of space buildings to visit or play any space entertainment facilities safely, e.g. space vehicle.

So, they must need to ceck anywhere the space planet's places to be ensured safe to build any buildings. Then, they can choose the suitable locations to build space entertainment facilities or buildings more confidently.

In fact, any space entertainment facilities, e.g. space hotels, space farming lands as well as space transportation tools, e.g. spce vehicle, spacecraft , these things will be value to be concerned to any space tourism leisure providers and it is business moral ethic responsibility to every one of them, when they plan to develop their space tourism business in any planets.

(3) Space tourism market competition challenge aspect

Thirdly, any provate space tourism development leisure businesses will face market competitive challenge, such as large spacefaring countries, e.g. US, UK have possible to dominate future space tourism leisure business (government can own space tourism leisure business). They will be main actors in space were nation-states. Large spacefaring counties can build the space vehicles, that can take people and cargo into orbit and to the Moon, or Mars crafted international space law and shaped the main investments in space tourism leisure technology.

So, it is possible that the own space technological developed countries, such as US, UK, these countries governemts will have possible to operate public fund to support space tourism leisure business. It implies that private space tourism leisure businesses will face public space tourism leisure business and themselve private space tourism leisure business market competition in space tourism leisure industry.

If these two countries governments also participate this private space tourism leisure market. It will raise market threats to any private space tourism organizations.

Whether will developed countries governments participate private space tourism market? It is possible that new commercial actors began to enter the space tourism leisure industry, looking to disrupt both space launch services ans use space in new exotic ways. For example, the US government also moved its purposeful degradatoin of the global positioning system (GPS), so US government will have effort to dominate GPS global positioning system communication business also. As this GPS communication business case, future US government has possible to decide to participate space tourism leisure business also.

However, in the future, space tourism leisure industry may contribute even more the developed countries, e.g. American, England economy. Space tourism and resource recovery, e.g. mining on planet, Moons and asteroids in particular may become large parts of that space tourism industry if these countries governments participated to this space tourism industry development. Of course, their viability rests on a range of factors, including costs , future regulation, international market competivitive problems and assumption about space technological development. However, these is increasing optimism in these areas of economic production to bring human space tourism leisure enjoyment and space mining resource development benefits. But the space economy is not just about what happens in orbits or how that alters life on the ground. The growth of this economy can also contribite to new innovations across all future possible unpredictable or undiscovered technological development, instead of space tourism leisure or space mining resource exploitation development.

Consequently, any space development technological governments will have possible to bring economic benefits from either only private space tourism leisure organizations or governments and private space tourism leisure both organizations cooperate to participate to achieve space tourism misson to contribute to global economic development and create new jobs to be employed in space labor supply market.

- Can space tourism business bring economy benefits

It is fact that space tourism activities have a positive and beneficial impact on eveyday life and society and this help space travelers to understand that, despite the high space ticket prices of any space tourism leisure choices. However, space tourism will bring scientific knowledge and technological knowhow and jobs to bring humn tangible or untangible both benefits. I shall indicate these benefits as below:

Although, space tourism leisure seems only leisure activities to be consumed to satisfy any space tourism individual travelling need. However, it can assign space scientists to research and attempt discovery these intangible benefits: Such as tele-communications revolution, satellite weather forecasting, mapping mineral exploration, water resource management diaster mitigation, national security or other undiscovered untangible benefits. Because every spacecraft needs to plan to fly to space, and it will reach any space planet stations, e.g. Mars, Moon planet when it visits these any one planet, the space scientists can attempt to find new undiscovered space resource , e.g. mining or finding new undiscovered satellite weather forecasting method when they can reach these planets to attempt to do space scientifical investigtion to research new space resource , or find any space stones attack to our methods to avoid earth disaster occurrence (national security mission), instead of the spacecraft catchs space passengers to visit these planets to enjoy these planets space entertainment facilities in their space trip journeys.

(1) On space resource benefit aspect

Hence, the space tourism intangible benefits include: space exploration and international cooperation is developing sophisticted space technologies by nations. For example, the images of distant stars and glaxies using Hubble telescope, research laboratory such as international space station to conduct experiments in biology, human biology, physics, Astronomy and meteorology under microgravity environment and testing of the spacecraft systems will be required for space tourism missions to the Moon and Mars. In the future, human would be able to have unlimited and clean solar energy from space for our industries as well as heating and lighting our homes. In the near future , it would be possible to disposed-off our nuclear waste safely and unexpensively and released towards the sun using a space elevator. We many become a space tourist in earth orbit or on the Moon or Mars. We may carry and extra-terrestial mining and even introduce the development of a multi-planet economy.

(2) On education benefit aspect

Another on education benefit aspect, space tourism can let space travelers to feel actual space learning experiences, during the spacecraft is flying in the space. Their space environment learning experience can include, for example: How many spacecraft have been launched by a given country? How many phone calls are made over a satellite? How many lives could be saved by resue satellites? How they feel differences when they are living in one space hotels, they are swimming in the swimming pools, they are visiting the space garden, they are running in one space sport centers, they are visiting in one space farming land, they are sitting or driving one space vehicle on planet land, or they are catching one spacecraft.

These space learning experience will let they feel what the actual differences between space environment and earth environment. It is one humankind learning experience education service in any space planet's Moon or Mars remote areas, bringing information and tourism entertainment facilities to the masses. The space experience learning knowledge can provide data to let these space travelers to know, such as how ships can be safe at sea, monitoring the threat of pollution, how enhancing durable medical instruments for better health-care enabling hikers and skiers to be located when lost, many more. So, it seems space tourism can bring much positive benefits as no negative impact on space activitied has been found by the society , the investments are made by the nations on space activites are justified and not the waste of money.

- What are the tangible social and economic benefits brought from space tourism?

In most advanced economies space tourism or space resource exploitation industry is seen as an enabler that improves lives and helps to develop both economic and social spheres. Space industry economic can include these aspect: Application of space technology to space tourism navigation, meteorological forcasting and broadcast of on live television and internet connectivity to lesser-known applications, such as precision agriculture, transport, tracking, resource extraction and monitoring of utility networks.

Additional application exists in the disaster monitoring and relif, insurance and military applications. Thus, data coming from satellites is important to all economic sectors, making the world a better and safer place.

International space tourism experience would suggest that space travelling leisure businesses deliver value by providing a central point for academia industry , defence and foreign entities to collaborate among themselves and with government and to facilitate the flow of knowledge and capital.

How can space tourism industry maximize the socio-economic benefits? In fact, our growing use of space derived data and systems is our growing dependence on a better and safer sapce planet, e.g. Moon or Mars and to provide space tourism safe services that space travelling service that space traveler all benefit from industry in telecommunication , health, transport , banking , security and climate change monitoring.

The space tourism positive influence result is long term, the positive contribution to our quality of life is real. In other word, the world for space tourism leisure activities is changing the internationally space tourism sector is experiencing a profound revolution.

In conclusion, space tourism leisure countries with historical leadership in space tourism have been under positive as a result of a tough financial environment leading to the definition of their space travelling technology priorities. In the meantime, new space entertainment travelling leaders, such as US, UK , even China, India have ambitions in space tourism through massive investments in the development of their capabilities in space travelling leisure business aspect.

So, the future space travelling entertainment market is large, due to China and India both have many rich people and high income people, who expect to consume in space tourism leisure trip at least one time in their lifes. Consequently, worldwide space tourism entertainment industry players are rethinking their busines models and strategies as they experience discuptive innovations, competitive space tourism entertainment and new drivers impacting the spacecraft and any space entertainment facilities manufacturing on Moon or Mars planet, launch and space tourism entertainment related businesses. Thus, we can in fact in talk about a new space tourism business, in which more and more innovative applications of space tourism data are developed dependence on space tourism data in everyday life rises and increasing share of economic growth relies on the space tourism market both in terms of opportunity benefits , e.g. India and China spce tourism potential market development and any concern space tourism job creation to every countries. Hence, space tourism development can bring positive economic benefits to any countries.

Space flight safe factor

To operate one space flight exploration organization, it needs to concern human safe flight factor. I shall indicate it needs to have these three stages to further develop its space exploration to continue to improve its safe space flight for every time of space flight.

Human future space flight missions will include these three stages to continue journey into space. The first stage is short term, NASA's return to flight after the Columbia accident. The second stage is mid term. What is needed to continue flying the shuttle fleet until a replacement means for human access to space and for other shuttle capabilities is available, and the third stage is long term, future directions for the kinds in space. Therefore, the space exploration organization can arrange the three stages to carry out any future space exploration activities. I believe it can improve every time of space flight more safe because it can ensure its space rocket engineering can be improved to raise safe level to let space people to catch to leave our Earth. However, any human future space flight, which must be enhanced safety of flight when carry on any experimenting space flight exploration missions. Because NASA's safety performance is a very important factor to influence any space people confidence to catch every sky rocket to leave our Earth to do any space exploration activities. So, eliminating and catching rocket risks will be any beginning and end than during the middle of any space flight exploration journeys.

Space people's life is the most important assets of any space exploration journeys. Because of the dangers of ascent and re-entry, because of unknown space environment and because we are still relative new comers, operation of shuttle and indeed all human space flight must be viewed as a development activity.

Thus, any every time space flight exploration missions will need to encourage to invent new space transportation engines (machine) or fuel, e.g. nuclear fuel to reduce the any space exploration journey accident risks and achieves to spend the fastest time to arrive any new space exploration destination. Thus, I believe any new space exploration flight will improve the space transportation technology and invent more new fuel and new space rocket manufacturing materials for future human any unknown space exploration flight demand. The three stages of improving space transportation include as below:

The beginning stage, for example, the space shuttle is as somehow comparable to civil or military air transport. They are not comparable; the inherent risks of spaceflight are serious higher. The recognition of human spaceflight as a developmental activity requires a shift in focus from operations and meeting schedules to a concern for the risks involves. Thus, the space transportation tools will be improved to protect space passengers safety: the improving the ability to tolerate it, repairing the damage on a timely basis, reducing unforeseen events from the loss of crew and vehicle, exploring all options for survival, such as provisions for crew escape systems and safe havens , barring unwarranted departures from design standards and adjusting standards only under the most safety-driven process.

The mid-term stage, the present shuttle is not very safe to fly in space. Thus, focus on safe return to flight is very important to every space flight journey rules , they leave Earth and arrive any another new planet destination, then come back our Earth again in every space exploration journey (flight). Thus, the energy will be space transportation tool one important factor. If the space transportation tool has enough supply, which won't stay in space and can not fly in space suddenly. Thus, the every time of the human space flight will be taken more time and effort then would be reasonable to expect prior to return to flight. Thus, human space exploration organization needs have higher reliability organization structure to manage every space flight, e.g. one is separating technical authority from the function of managing schedules and cost. Another is an independent safety and mission assurance organization.

It is the capability for effective systems integration perhaps even more challenging than these organizational changes are the cultural changes requires. Thus, the cultural to safe and effective space rocket operations are real and substantial. If the space exploration organization has good culture to let every staffs can communicate easily. I believe the every time space exploration accident will be reduced. Examples include: the tendency to keep knowledge of problems contained within a center or program, technical decisions, without in -depth, peer-reviewed technical analysis, and an unofficial hierarchy or system created by placing excessive power in one office. Such factors interfere with open communication, the shared of lesson learned, cause duplication and expenditure of resources and create a burden for managers to reduce undesirable characteristics threaten safety.

Thus, any space exploration trip, rocket equipment safety and check are very important factor to prepare for every time space flight. The reason is that space flight must guarantee any space people who can come back Earth, if the rocket equipment are poor and lack maintenance. The, the space people whose life is dangerous. Due any space exploration organization mission require human presence in space. For example, president John Kennedy's 1961 charge to send Americans to the moon and return then safely to Earth. Thus, the space exploration organization has attempted to carry out a similar high priority mission that would justify the expenditure of resources on a scale equivalent to those allocated for project Apollo. Also, the space exploration organization has had to participate in the give and take of the normal political process in order to obtain the resources needed to carry out its programs.

Another main successful factor in the final stage, the space exploration organization needs have a clearly defined long term space mission to commit over the past decade to improve future space exploration flight safety by developing a second generation space transportation system. So, for long term, the space exploration organization should need to plan for future space transportation capabilities without making them dependent on technological breakthroughs.

For example, mission for a post Apollo effort that involved full development of low-Earth orbit, permanent outposts on the moon, and initial journeys to Mars planet. Since that rejection, these objective, have reappeared as central elements in many proposals, setting a long term vision for any space exploration flight programs in the future.

Thus, space organization future space exploration mission for 21 St century is to lead the exploration and development of the space frontier, advance science, technology and enterprise and building institutions and systems that make accessible vast new resources and support human settlements beyond Earth orbit from the highland of the Moon to the plains of Mars. Thus, the space exploration organization limit is to conduct the research required to plan missions to Mars and/or other distant destinations. This is the most safe space flight distance limit by the space rocket equipment, machine installation , quality and effort to guarantee space people life safety when who catch the space rocket life safety when who catch the rocket to leave Earth to arrive any space destination in any space flight. However, human travel to destinations beyond Earth orbit has not been adopted because it is too far space flight to cause accident risk. Hence, space exploration organization future invention of long term need is that the role of new space transportation capabilities in enabling whatever space goals need to choose to pursue for human present in Earth orbit vision.

In conclusion, space exploration organization needs to in-depth examination space shuttle safe issue, how to reach an inescapable design of the space shuttle, because that the design was based in many aspects on how absolute technologies and because the space shutter is now an aging system , but still developmental in character, it is in the space organization is interest to replace the shuttle as soon as possible as the primary aim for transporting humans to and from Earth orbit.

- Space exploration organization mission and strategy

Space exploration organization communication strategy

I recommend any space exploration organization needs to the message concerns how the role of humans are actual physical presence in space exploration missions succeed. Because the positive message will give good idea of space exploration and then design and build means to carry out right space exploration direction to let humans to know whether any space exploration missions' goals, objectives and what humans benefits (welfares) who can earn.

The message includes such as these primary role of humans, therefore, is to provide the inspiration and create the vision which produces the motivation in those who then go on to make it a reality, e.g. the space exploration mission is to bring their human intellectual capability to bear in designing the technical systems required for space transportation and devising the scientific experiments associated with space exploration from its beginnings.

Thus, any space exploration organization needs to let humans to know whether what benefits humans will earn after it carries out any space exploration experiments possibly. I believe that the exploration of the Earth's great expanse (the sea, the undersea world, air and land) is the ultimate role played by humans in body and in mind, and apply their intelligence, emotions and most importantly of all, their superior cognitive performance. So, this is the role now played by astronauts, explorers in the true sense of the world.

Why does space exploration organization need to be the role of communicator? The reason is because there is the role that space organization's need to play as communicators, journalists or other communication professional. It is they who provide the link between those involved in the project and taxpayer, who are entitled to be informed about the fascinating news on space.

Moreover, space exploration organization staffs need to give message to let humans to know why these playing roles are entirely human specific and can not be fulfilled by machines. For example, roles prior to human intervention, such as accompanying humans and performing tasks, which are repetitive and unpleasant out satellites too high a risk. By sending out satellites to explore our solar system humans have already begun to explore universe into reality Robots. On the other hand, may be things, but they are not visionaries and nor are they inventors or explorers. Any achievement they accomplish are in fact space organization staffs who designed and programmed them. Also, humans remain the best available cognitive machine in any environment that may be subject to significant variations relative to the model initially made of it. Thus, space exploration organization needs to explain, such as why in the general context of space exploration, even of most missions are robotic, remains technology challenges, it presents push engineers to the very limits of what can be achieved.

In the future, humans will earn these benefits or from any space explorations new invention possibly, such as fuel cells, the microcomputer, high performance materials, medical advances, new management techniques for major projects, quality and reliability control in industry etc. The most important space exploration organization needs to positive message to let these groups of people to human what which is doing in our societies. Then, which will cause different actors to become involved from thinkers, visionaries and inspirational figures in the form of writers and film makers to scientists, engineers, philosophers, politicians, economists, physicians, journalists, authors, space travelers (astronauts), but also adults and children space story book readers alike. Thus, space exploration organization is truly multi disciplinary enterprise. Moreover, in the present day, normal escapes being concerned by space, as much due to its contribution to daily life and the knowledge it beings of the Solar system and the universe. It seems space exploration organization will influence human past history will be changed to develop. Whether it brings positive or negative change. The space organization must have responsibility to keep its any space exploration missions leader position in our Earth. It implies it is also one social responsible organization for future global human benefit (welfare).

Also space exploration organization needs to let humans know what it's future aims (intentions) are to let humans know whether why it plan to implement. Such as it needs to choose destination has typically been the Moon, it had increasingly come to focus its attention on Mars and even further afraid. Moreover, it also needs to know humans to know the modes of future space transportation which described have tended to be those of the period concerned: ships, horses, birds, balloons, canons, rockets and even others of a more esoteric nature, even solar sail or nuclear soil further space transportation technology development. In addition, space exploration organization can need to describe where are further orbital space stations in space different locations and explained the various applications of satellites and spacecrafts to let human to know clearly.

Even, space exploration organization also need to let humans to know what are their technical challenges, it will encounter in any space exploration stages to let humans to know. Although, the complexity and changer involved in spaceflight is such for a long time to come there will be a need for experts, whose focus by necessity. So, the general public will know or recognize why it's technical challenges will cause and how it will attempt to solve these technical challenges. It aims to let humans to ensure more than 40 years of spaceflight, the adventive of space, which for technical reasons is inevitably reserved to a " happy few", remains very much the preserve of specialists, cooperation to research how to solve any technical challenges to achieve success in any space exploration mission consequently.

- Space exploration organization team leaders and their teams

The first team members are program chiefs and mission message are request to the be backroom generals with a great many human qualities. They must having to achieve great technical exploits and manage their teams with care when at the same time ensuring they deliver in timing and one budget. Even the very best robot-machines and computers available are of no help to them in coming up with the initial idea and architecture for their systems. Indeed, in that initial stages, some program chiefs, even insist on their management team using only paper and pencil writing. Once the concept has been defined, they then need computers to speed up and develop the project.

When these leader figures are fortunate enough to see their program in orbit and crowned with success, their experience and methods can be of use, to equally computer technical sectors. They can also be passed on to following generations, thus safeguarding, for reasons of economics and security, the know how acquired by their teams. Another team members are scientists and those responsible for the technical side of program are not generally skilled communicators by nature, those with communication to public , such " communicators" could be awarded special prizes. Communication on the space sector can't be left to " communication specialists". Otherwise, there is a risk, it will be perceived to be doomed to failure.

Space exploration organization education is such as strategy space exploration organization communicator, are there to inform, the teaching profession for its part, must perform a vital education role, helping people understand the universe in which they live. Space exploration represents a unique opportunity to explain the situation of our planet within the solar system, asking questions such as: How does the sun function? What are the origins of the Moon? Why does Venus have such a pronounced greenhouse effect? Is there or has there ever been life on Mars? Do asteroids pose a serious threat? There are all questions which today, our schools don't even attempt to answer. Thus, space exploration organization can be one educator role, instead of space explorer role.

Thus, space exploration organization has mission to assist universities to promote space exploration education knowledges. It brings this question: What other technological and scientific program is better equipped to meet these objective than space exploration , with its crewed emissions component. So crucial to the promotion of a European industry, so visible to the general public and so efficient in inducing younger generations to take up scientific and technical careers? Thus, the space education courses can include space exploration industrial applications, a new area of investigation to scientific fields, fundamental physics, cellular and vegetal biomedical research and human and animal physiological research etc. subjects. For example, teaching how to go to Mars or other planets and manage to live these will require a knowledge of how to energy in innovative ways for the purposes of managing electricity generation requirement will be to learn how to manage scare resources in an efficient way (air, water and waste recycling). So, teaching of progress will have to be made in advanced robotics in particular in the area of effective and human robot interaction.

In conclusion, all these space exploration science education knowledge will be important to be taught to let younger to pursue space resource exploration dream for human future live

- Space exploration organization's

Human space life science factor

What is human space life science strategy?

One space exploration organization needs have good human resource strategy to implement every space exploration mission. Critical to this expansion of human presence in space science will enable mission success by focusing on risk reduction and optimizing astronaut health an productivity through space organization's human-centered science, operations and engineering core capabilities.

Thus, the space life science strategy's strategical goals, and objectives were developed on the basis of a situational analysis conducted by key members of the space life science civil service and contractor community, and are consistent with agency goals and scenarios for the future.

This strategy mission is to optimize human health and productivity for space exploration and its vision is to become the recognized world leader in human health, performance and productivity for space exploration . It's strategic goal are aimed at driving innovations in health and human system integration, adapting its portfolio and strategies to the changing environment and creating enthusiasm for space exploration through education. Also, the space life sciences human strategy aims to achieve every space exploration research more success, more efficient, focuses on client (human) needs and facilities communication of risk to public and the value of space life sciences to its stakeholders (governments, universities, societies).

How can human space life science strategy implement?

The space exploration organization needs to be dependent upon healthy, productive astronauts to achieve mission success. Thus, space people health are very important factor to influence their every time space flight in success. If the space people have unhealthy bodies , which will influence whose work performance and every time space exploration mission can't finish easily. Thus, the human space life science strategy needs to ensure every space person has health body to work efficiently and reduce whose death or accident risk when they are working in space environment, due to space environment is one strange bad color environment and it is very difference to our Earth environment to unsafe to work by these factors:

such as, it's temperature is low, cold and no air or oxygen to be supplied to let human to breathe and it has unknown diseases in space. Thus, they will face any life danger when are working in space environment. If space organization lacks one human space life science strategy to help them to fight any unknown attack from space environment. The, they are very dangerous to attempt to catch space rockets to leave our Earth to do any space exploration activities.

However, the space life science strategy can divide these three timeframes consistent with

- Near –term (1-5 years)
- Mid-term (6-10 years)
- Long-term (11-20 years)

The space life science strategy mission is that optimize human health and productivity for space exploration . Thus, all space life sciences human health and countermeasures research, medical operations, habitability and environmental factors activities, and directorate support functions are ultimately aimed at achieving this mission. Their activities enable mission success, optimizing human health and productivity in space before, during and after the actual space flight experience of their flight crews, and include support for ground-based functions.

The space life science strategy vision is to become the recognized world leader in human health, performance and productivity for space exploration. Thus, to achieve the vision for space exploration , they must drive human health, performance and productivity innovations, adapting their strategy to the changing environment. To do this, the space exploration organization needs have a future scenario for space life science strategy such as below:

- Future core capabilities will include the expertise to address space medicine, the physiological and behavioral effects of space flight, space environment definition and space human factors.
- Research plans are on the basis of a standard –based risk mitigation approach to ensure goals are achieved.
- Civil servants will balance delivery of health and performance services and focused research and technology development with smart buyer and management expertise to integrate space life sciences efforts.
- Strategy relationships will be utilized to achieve the full complement of space life sciences core capabilities necessary to achieve vision and enable mission success.
- Space life science strategy will transition from being a managing partner to a contributing partner, arranging the resources and innovations of other

organizations to meet specific exploration needs, e.g. universities, government or business biomedicine organizations.

- Operations will effectively transition the space people skills and facilities from shuttle and assess and engage in additional government and commercial space flight operations opportunities where appropriate.
- An expanded client base that may include additional international and academic partners , as well as commercial alliances.

Situation analysis

A situation analysis was conducted to determine its mission and to identify the factors most likely to influence its strategy development and affect achievement of its goals and objectives . It will trend to concern life sciences and space flight of internal and external environments. It needs to image these assumptions to decide its situation analysis as below:

Thus, the first assumption is that it needs to assume that human will continue to be an important component of the vision for space exploration, and as a result there will be an ongoing need for space life sciences core capabilities, including human-centered science, operations and engineering to mitigate the health and performance risk of human space flight.

Another assumption is in the longer term, there will be a greater focus on crew autonomy and increased human-robotics interaction as mission durations increase and are extended to travel to and on planets, and as a result, these is a continued need for research and development activity focused need for research and development activity focused on exploration risk reduction.

The next assumption is the pace of biomedical change will continue to be more rapid in external versus internal environments. Thus, solutions are more kinds of likely to be developed external to fight any different new unknown new diseases to attack to influence space people health to be poor , even cause death in possible.

What are space life science strategy goals?

On health innovation hands, the space exploration organization will drive advances in medical and environmental health for space flight in order to meet established space life standard and mission needs. Thus , innovation on medicine and biomedical / environment technology and processes will be developed, implement and incorporated into mission achievement.

On education hand, it needs to train in multidisciplinary life sciences, experts in exploration life science and that this is a continuous infusion of space ,life science into the public , government , academic and commercial sectors. Thus, the space life science education aim includes to teach the human system risk management, strategic relationship of any space missions, future space business model and space communication strategies.

The goal-specific strategies and measurable objectives can be developed for years 1 to 5 years It's objectives can include: optimize internal core capabilities throughout the planning cycle to enable the vision for space exploration with budgetary constraints, establish strategic relationship to achieve the full complement of life sciences capabilities necessary to be best in class , establish a center to integrate human health and performance efforts and expertise for space exploration worldwide, implement an internal and external communication plan to increase the life sciences value to encourage space human life education development for long term in commercial space flight sector.

Health innovation goal

Thus, one space exploration organization whose health innovation strategy is the main factor to influence its any overall space exploration missions inn success. Thus, it must need to spend more money and time and resource to ensure its health innovation implement can be succeed to reduce further every time space people's mission of physical illness or death or accident which are caused by space diseases . Thus, it will drive advances in medical and environmental health for space flight in order to meet establish space flight health standards and mission needs. Also, it needs to attempt to do any biomedical experiments to avoid space people who can contact to cause illness from any undiscovered space diseases.

Hence, the invention of space medicine, biomedical / space environmental technology and processes will be developed, implemented often every day/ IT needs to seek or gather every time practical space environment biomedical existing data and knowledge as a base for launching health technologies and to revise every time space biomedical experiment failure to find failure reasons to achieve the most absolute discovered any space unknown diseases biomedical experiment results. Thus, the improved methods and practice or recording data must concern to goals for human space exploration, attempting towards data gathering top continuing to achieve the best levels of evidence for answering operational and clinical questions regarding human health, safety and performance, during space flight and exploration and an evidence-based risk management approach to prioritize tasks.

The space biomedical experiments data gathering can consider human factors engineering, habitability design and human-robotics interaction will be recorded to analyze experiment result every time. These results will be developed, implemented and incorporated into mission architecture solutions to address the human as an element of the overall space system.

● Prediction on future trends in human space flight and future space human life science strategy relationship.

In the future, the relationship between future trends in human space flight and future space human life science will be more close. These reasons are that the trends in terrestrial life sciences will save as change drivers for space life sciences, include advanced in nano health, genetics, biocybernetics, self-constructing materials, human computer interfaces, medical and pharmaco-therapeutics, multi-scale physiological modelling and other biomedical technologies.

In conclusion, due to space exploration organization's objective is low tolerance for a risk and emphasis on risk quantification and reduction activities. Thus, the space human life science

strategy will be one important factors to cause any one space exploration organization's any missions in success.

● Why does Japan space organization consider space human life science?

Japan has acquired and advanced various space technologies. Through, these technologies level to allow to play a core role in the international human space activities . However, it's space exploration success is due to it concerns to achieve its space human life science strategy for its main point.

What social benefit from its utilization of the space environment to Japan. Because it concerns how to protect space human life during who are working in space. Thus, it can bring more social benefits to develop its space exploration industry for long term as below:

● Because it's space people can have health bodied, so who can attempt to any space science exploration experiments in space

environment as well as space human life science can raise Japan space people confidence to attempt to do every time space exploration activities in space environment. Consequently, they have confidence to catch rockets to go to space to gather various resource to do any space exploration to get research results more easily, which were achieved through utilization , such as micro gravity environment, that could not be produced on the ground, these outcomes include: protein crystal growth, that may lead to the development of new drugs, materials creation for next-generation semi-conductors, and establishment of the technology for cubesats deployment, etc.

● Due to Japan space exploration organization concerns space people health issue. Thus, it has manned space flight capability can conduct youth development activities , with their own astronauts and such astronaut-led activities have aroused the younger generation's interest in outer space, taught them the importance of making efforts to making of one health space scientists confidence to pursue this space exploration industry.

Hence, when all Japan space scientists who own health bodies, then who can be one expansion of humankind's space of activities in this area create knowledge of planetary science and the quest of health space life and also contributes to the increase and accumulation of intellectual assets of all human beings.

The most reason of Japan's belief of space human life science strategy is very important , because it needs to prove human can live in space environment. Thus, if Japan's space scientists can have health bodies to do any space exploration experiments, then who is still health to go back Earth. Then , it proves the life support technologies , the space environment and health management and the maximum energy conservation. This leads to the enhancement of corporate brands and international appeal of technical capabilities , and is directly linked to resulting problems Japan faces , such as its aging population and lack of natural resources.

In conclusion, space human life science will influence Japan space exploration industry more success. Otherwise, if it chooses not to implement this space human life strategy. It won't have enough health space scientists to attempt to catch rockets to go to space to do any space exploration experiments more success in long term, e.g. seeking Earth another planets to provide Japan people to live, raising Japan young space scientists confidence to attempt to go to space to do any experiments because the Japan space exploration organization can provide new bio medical invention to supply when they are catching in the space rockets. If they feel that they are comfortable, they can eat or drink the new bio medical invention to avoid the space disease attack to cause their death or physical illness threat.

In conclusion, space human life science is very important factor to influence future every time human space exploration mission successfully.

Reference

Cater Iain , Carl 2010, " Steps to space: Opportunities for astro tourism development, tourism management 31 (2010); pp. 838-845; Elsevier Ltd, DOI: 10:1016/j.tourman. 2009.09.001

Collins Patric, Iwasaki Yoichi, Kanayama Hideki, Ohnuki Misuzo 1994, comercial implications of market research on space tourism. journal space technology and sciences , vol. 10 no 2, 94 Autumn, pp.3-11. copyright: Japanese rocket society; available at: www.spacefuture.com/archive/commercial-implications-of market-research-on-space- tourism.shtml.

Collins Patric, Marita M; Stockmans R. and Kobayahi S. 1996. "Demand for space tourism in America and Japan and its implications for future space activities ". sixth international space conference of Pacific basic societies; Marina del rey; California: Advantages in the Astronautica science (AAS paper no AAS 95-605) vol. 91. pp. 601-610. Available at: http://m.internationalaerospaceconsulting.org/upload/space %20Future%20-%20Demand% 20for%20space%20Tourism%20in%20America%20Japan.pdf

ESA 2008, " Richard Garriott, millionaire American space tourist. blasks off of international space station". published in 12.11.2008. Huffington post, seen on i01.04.2015; available at: http://www.huffington.com/2008/10/12/richard-garriott-milliona-n-1333940.html.

Klemm, G., & Markkanen, S. (2011). IN A Papathanassis (ed.) The long Tai , tourism (pp.95-103). Weisbaden, Germany : Gabler Verlag; Springer Fachmedien Weiesbaden GmbH.

KSCVC Visitors, 2013; MRI 2013 Market by Market

Von Der Dunk , F. (2012). The integrated approach. Regulating private human spaceflight as space activity, aircraft operation, and high-risk adventure tourism. Acta Astronautica, 92(2), 199-208.

Development space tourism successful factors-space boat creatve technology

Space tourism passengers leisure feeling

In future tourism leisure development, if we hope to develop our future space tourism entertainment in success, we need to know how to persuade general rich people how to make decision to choose to catch which kinds of rockets to travel outer space, because many of they will feel space is horror when they need to stay at the strange dark low degree temperature and none heavy weight body body feeling environment space environment. When they catch the rocket to visit the dark space strange environment.

In common, any space rocket passengers may usually feel dangerous , when their space rocket is crashed by any space stones suddenly. If one big size space stone flys toward to the space rocket direction to crash suddenly, due to the space rocket can not predict when any space stones will fly toward to this space rocket to crash it suddenly. Hence, space strange dark environment may influence many space tourism passengers feel fear to catch space rocket to fly in space environment in long time.

However, although flying to space environment toursim which may bring exciting and pleasure and enjoyment feeling to any one when they are catching the space rocket to fly to future space station, moon, even fly to space hotel to live one night or more, or stay the rocket to view our earth in different space environment locations for several minutes etc. different kinds of space tourism journeys. But, when the space rocket stays longer time in space environment, then it may bring more dangerous to the space rocket and the space passenger individual life both. Because many different kinds of space accidents may occur, e.g. space rocket has fire happens, machine equipment is damaged, space stones crash the rocket, space steel rubblishs crash the roctket, even space rocket windows may be damaged by any things etc. So, different kinds of accidents may be difficult to predict. Also, it implies that we may ensure any kinds of space rockets have chance to occur accidents when they are staying in the strange and dark space environment in any time.

Consequently, it will bring fear feeling to let any one space tourism passenger feel afraid to catch the space rocket to travel space long time. In special, if the space tourist is very rich, he must feel his life is value, he ought not lose his life due to this space rocket is crashed by any space stones, or space rubblishd ot itself damage factor etc. different space accidents occurrence. Hence, if any one space tourism service provider hopes that it can encourage many space tourists choose to catch itself space rocket in preference. It must need to let many its space rocket passengers feel itself space rocket is more safer or brings more safe feeling to let its all space rocket passengers to feel. Moreover, its needs to let many space tourism passengers may feel its space rocket is more safe to compare other space tourism leisure service providers their space rockets facilities.

Consequently, there are many space tourists may be influenced to choose to catch this space tourism lesiure service provider's space rocket to fly to space, because if they feel other space tourism leisure service providers , their space rockets are not safe to compare this space tourism leisure service provider's space rocket. Then, this space tourism leisure service provider may design different kinds of unique space journeys in order to let any space tourist feel more safe to stay longer time in space environment. Hence, it may help any one space leisure tourism service provider raise to charge higher ticket fee when its every space journey is longer time to compare general short time space journeys.

Is right time to develop space tourism lesiure

Nowadays, human can have different kinds of lesiure choices. They may include tourism in ourselves earth, sports, visiting cinemas to see movies, listening musics, reading ebooks or paper books, visitng theatre to see art performances, seeing paints etc. any every different kinds of lesiure activities. However, comparing among of any kinds of general lesisures, tourism lesiure must be the most expensive kind of entertainment activity to compare all of above these different kinds of lesiure activities. Hence, it brings this question: Is right time to develop space tourism leisure market?

In fact, future human travel choice has only these both kinds: One is travelling in ourselves earth, another is catching rocket to travel space, e.g. visiting space station, staying on space one location to view ourselves earth or moon, or visiting moon, living space hotel one night or more etc. different kinds of space tourism leisure activities. However, I beleive that it is right time to develop space tourism leisure. I shall explain as below:

The first reason, all of us ought know that this kind of disease COVID 19 had influenced global many travellers had began to feel fear to catch air planes to go to other countries to travel because many travellers began to believe that they may get this kind of illness easily when any one of air plane passengetr is sitting close to the COVID 19 patient, even any one air plane passenger may also be gotten by air when any one COVID 19 patient is sitting in the close window air plane seat. Hence, COVID 19 disease is significant to influence global many travellers began to reduce to catch any air planes to go to any countries to travel frequently, even the having travelling habit travellers will choose not travelling any countries any times by air planes.

So, it seems that COVID 19 disease will have long time to influence futuer global tourism leisure market development in our earth. It is one negative psychological factor of " feeling afraid to catch air plane" to any one liking tourism leisure tourist. So, when one likes often to go to other countries to travel, in special, the high income or rich people group, when they feel fear to need to catch air planes to go to other countries to travel, they must feel bore, due to they must only stay in themselves home countries by COVID 19 disease attacks. In fact, although these high income or rich people tourist group can choose any kinds of leisures to replace tourism leisure activity, e.g. sport, going to cinemas to see movies, going to concert halls to listen musics or songs, going to theatres to see art performances. But, all of these different kinds of entertainment activities can not satisfy themselves leisure psychological needs, because they have much money, they like to spend money to go to any countries to travel for one week, or two weeks , even one month more in their holidays.

Nowadays, because this kind of COVID 19 disease influences they feel afraid to catch air planes to travel to any countries frequently. They avoid to catch air planes to contact any one, he/she may have COVID 19 diease, even when they arrive the traveling destination, they none COVID 19 disease travellers , they feel that their bodies may be goten COVID 19 disease, when they need visit any one hotel to rent rooms to live, or rent appartments to live, they must need to live strange hotel room. However, the hotel room may be lived by any one COVID 19 patient, when the tourist choose to live the hotel room, he will have chance to get COVID 19 disease due to he contact any things in the hotel room. Even, any one tourist needs to catch any train, bus, tram, taxi, ferry , underground train etc. different kinds of public transport tools, they may have chance to get COVID 19 disease when they are catching the public transport tool, if it has any one COVID 19 disease passenger is catching the same bus or other kinds of public transport tool, or they need to go to any one restaurants to eat breakfasts, lunchs, or dinners, they need to sit the table with the COVID 19 patient together, ot they need to walk on the streets or they need to climb mountains or they need to visit any travelling destinations etc. to do any kinds of travelling leisure activities, that they need to do in their travelling jounrneys. So, it implies that they have chance to bring COVID 19 disease to themselves bodies due to they have chance to contact any one strange person in any travelling destinations. Hence, they will feel that they may be contacted to get COVID 19 disease , when they are staying in the travelling country.

Consequently, it seems that COVID 19 siease may be one important factor to influenc global any one feels afraid to catch air planes to go to anywhere to travel. SO, such as the special rich travellers group, they only have another travel choice to replace tourism in earth, it is " visiting space tourism". Although , every tecket for any one kind of space tourism , it may be very expensive, but for this special rich traveller groups, their space tourism leisure needs ought not be influenced reduce by expensive ticket factor easily. Because they feel that space tourism will be only another kind of expensive lesiure activity to replace tourism leisure in our earth.

The another reason is that global rich people number had been increasing every year significant. For China and US and UK and India etc. these high population countries, their rich people number had been climbed countinue every year . Hence, it implies that future many people have effort to epend at least one time of space tourism leisure, due to rich people number had been increasing in countinue. Moreover, many habit travellers began to feel that our earth has none any value destinations to visit, when they had visited any one travelling destinations at least one time. So, when many travellers feel that they had attempted to visit any vaue travelling destinations, then they won't like to buy air ticket to visit any one of these travelling destinations again. So, the only one tourism choice, is that visiting to space 's any destinations, e.g. visiting moon, visiting space stations to live one night hotel hotel, staying on space one location to vire ourselves earth or moon etc. different kinds of space journeys. Because may of rich travellers had felt that travelling in oursleves is low valur leisure and it can not attract any one rich tourist chooses this kind of lesiure again, as well as they had felr that earth has none any destinations can attract them to travel, so it is only space tourism may replace tourism in ourselves earth.

On conclusion, it is right time to prepare to invest another new kind of tourism leisure, such as space tourism , it may repalce traditional tourism in earth leisure. SO developing space tourism leisure is one value leisure developing market.

THREE

CREATIVE TECHNOLOGY (AI) PREDICTION CONSUMER BEHAVIOR TOOL

How can artificial intelligent tools predict consumer behavior in vehicle market

What is (AI) consumer behavioral prediction tool? How any why will (AI) tool assist manufactures to attempt to predict consumer behavior before and after consumption occurrence? First, I shall indicate how to apply (AI) tool to predict vehicle product consumer behavior case example.

Nowadays, many vehicle manufacturers hope their vehicles can attract to vehicle buyers to choose to buy their vehicles. However, there are many different brands of vehicles to provide to them to choose, so the vehicle market competition is very serious.

How to judge their different kinds of vehicle price which is reasonable acceptance to attract vehicle buyers to choose to buy the brand of vehicle manufacturers' any kinds of vehicles, e.g. fast speed sport style vehicles, comfortable and slow speed common cars, for four passengers common small size or more than four passengers common large car size? How to evaluate the vehicle prices issue is important factor to influence vehicle buyers' choices. Either if the brand of vehicle price is too high to compare brands, it will influence many vehicle buyers choose to buy other brands' vehicles or if the brand of vehicle price is too low, it will influence vehicle buyers feel this brand's vehicle's quality is worse to compare to other vehicle brands' similar vehicle products.

Thus, if the brand of vehicle manufacturers can predict how to design vehicles which can attract many vehicle buyers to choose to buy whose any vehicle products. What are future vehicle buyers' favorable vehicle styles? Then, the vehicle manufacturer can concentrate on manufacturing the kind style of vehicle products to sell already. It will reduce its vehicle manufacturing investment risk.

How to apply (AI) tools to predict vehicle buyers' behavioral consumption model? Whether artificial intelligent tools can predict automotive buyers' behavioral consumption model and predict future trend. In fact, automotive brands and dealerships are facing an increasingly competition when attempting to manually gathering the vast quantities of data required to create customer focused programs that increase retention, ultimately new sales and service automotive business. Building a based on that client's intrinsic needs and interests to any kinds of automotive vehicles at any given time. This is especially true in the automotive industry where the time span between purchases is measured in years. Because vehicle buyers would not like often to change their old vehicle to another new one. So, their decisions to buying another new vehicle, the time is usually after one year, even longer time. Hence, it seems any vehicles won't be frequent consumption products to the owned at least one vehicle family consumers (vehicle buyers).

Hence, how to predict vehicle consumers' taste or preferable which styles of vehicle choices issues is very important. If the vehicle manufacturers can not manufacture any attractive vehicles to sell easily in this year. Then, it will lose time, money in this year because it won't know when the owned least one vehicle users or non-owned any vehicle users who will decide to buy one new vehicle or change another new vehicle ensure. The different brand vehicle dealers will possible wait more than one year to attract them to buy their vehicles if their styles are not attractive to compare other brands of vehicle competitors.

However, artificial intelligence and machine learning can help any vehicle manufacturers to find solution to solve patterns in highly to solve patterns in highly complex data-sets that are beyond the capability of a human brain, and then building and automatically acting on the customer insights it generates.

Given the automotive customer need for individualized communications, this technology is positioned to become a critical component of any successful vehicle retailer's domestic or/and overseas vehicle markets. How can vehicle manufacturers and retailers use (AI) to enhance their vehicle marketing campaigns? How will (AI) affect their vehicle sale marketing strategy? What criteria would they use when selecting on (AI) solution?

Vehicle consumers today are able to quickly access different brands of vehicle information, research vehicle products and reviews, negotiate prices and compare one vehicle brand or retailer to another resulting of the brands of vehicle customers. At the same time, the rise of " big -data mining", wearable devices that track user's every move and preference and greater contextualization in advertising and social media has resulted in consumer expectations of individualized. Thus, it seems that (AI) tools can be used to gather " big-data" and then they can make human's mind to analyze how to design kinds of vehicles to satisfy vehicle buyers' needs.

As automotive vehicle marketers can apply (AI) tools to achieve messaging strategies to meet the needs of this new generation of informed vehicle consumers, using data from a variety of sources to move from a variety of sources to move from mass- messaging to more personalized messages aimed at particular vehicle buyer segments, e.g. fast speed sport vehicle buyer segment, slow speed comfortable small size or large size of buyer segment. However, when 90% of vehicle marketers believe having a single vehicle buyer view is important, only 6% have achieved it.

However, one of the main issues vehicle marketers facing is the lack of capacity to efficiently sift through and analyze the massive vehicle buyer amounts of data required to create vehicle buyer individualized vehicle customer experiences easily. This is especially difficult for automotive dealers, the long periods between purchase cycles, and the highly considered nature of the vehicle purchase means that each vehicle dealer needs to not only track a large number of potential vehicle customers for an extremely long period of time, but each of those vehicle customers will generate a huge amount of different kinds of vehicle behavioral consumption data as they research their next vehicle purchase. However, by choosing the right (AI) technological tools and programs , vehicle dealers can solve this big data gathering challenge into a major advantage.

For Forrester vehicle brand example, vehicle consumers have more power over the Forrester vehicle brand's reputation than ever before. Mayne, L. (2014) indicated that Forrester calls this new (AI) tools is the " age of the vehicle customer", a 20 year business cycle in which the most successful vehicle enterprises will reinvent themselves to systematically understand and serve increasingly powerful vehicle consumers. To win in this new age, Forrester declares companies must become vehicle customer obsessed and the only sustainable competitive advantage is knowledge and engagement with customers, such as (AI) gathering data knowledge.

Thus, the biggest challenge vehicle businesses currently face is not the collection of a large quantity of vehicle consumer data, but what to do with that data once they have it. Even at a large vehicle data research firm, the data sets are often too big for a single analyze, or even a team of analysts to sort through and draw conclusion from. However, enter artificial intelligence and machine learning , an efficient technology solution that can continuously find patterns in highly complex data sets that are way beyond the capacity of a human brain and then automatic drive action based on the customer insights is generated.

What is (AI) machine learning tool? Machine learning is a type of (AI) that learns from data and is not explicitly program. Think Amazon, face book. Machine learning serves up relevant content based on an individual vehicle purchase behavior and experiences. More simply, machine learning is a computer program that can learn relationships between data, subject those learnings to errors functions, and then learn from its errors. The program in effect, trains itself.

Lee, T. (2016) explained that "Thus, (AI) tools can learn deep a more advanced branch of machine learning inspired by how our brain's nervous function, has also been found to be especial effective in identifying patterns from data."

When this way sound is complicated from a vehicle dealer perspective, the implementation of a marketing program driven by artificial intelligence can take care of these tasks in an automatic vehicle fashion with little to no manual intervention required from the staff at time vehicle stores.

In practice at a vehicle dealership, the program will continue track vehicle customer behavior online, merging that data with any offline source (like CRM or DMS data) and then analyze this aggregated vehicle buyer data set to predict what vehicle customer may be shopping for and what information they might like to relevance from different kinds style of vehicle design photos.

Why can (AI) be applied to predict consumer behaviors?
Artificial intelligence refers to complex in vehicle market, machine learning that posses the same characteristics of human intelligence and that have all our sense, all our reason and think just like human do. Besides, machine learning is the practice of using algorithms to collect and examine data, learn from it, and then make a determination or prediction about something in the world.
The machine is " trained" using large amounts of data and algorithms that give it the ability to learn how to automatically perform a task with increasing accuracy. Otherwise, deep learning is primarily based on artificial neural networks inspired by our understanding of the biology of human's brains.
Deep learning breaks down tasks in ways that enables machines to assist us with increasingly complex tasks, driverless cars, better preventive healthcare and more accurate product recommendation (including vehicle recommendations). So, such as why (AI) technology can be applied to predict how vehicle consumer behavior changes to bring to judge whether vehicle consumer will like what kinds of vehicle styles next year. Then, vehicle manufacturers can gather overall vehicle consumer data to analyze and conclude the more accurate vehicle design direction for next year any new design vehicle manufacturing products.
Thus, (AI) machine learning can help vehicle manufacturers to solve how to design any new vehicle products challenge. A vehicle is both one of the most important and carefully considered purchases the majority of people will ever make in their lifetime. It is also a purchase that tends to be fundamentally tied to a person's identify and view of themselves. As the same time, vehicle consumers changing lifestyles result in changing vehicle needs, e.g. the young sport car enthusiast matures into the family driver.
Automotive dealers need to remember that vehicle customers and prospects are individual human beings with risk, complex and ever-changing lives factors, these factors will influence every vehicle consumer why who feels has vehicle purchase need, and how who choose to buy the first vehicle if who decided to buy the first vehicle.
The (AI) technological customer behavioral prediction tool seems to be the best vehicle salespeople in the world are those that know every one of their vehicle customers. Their likes and dislikes which style of vehicle design, preferences and changing tastes to vehicle choices. The capacity of the human brain, however, limits us from achieving this type of vehicle sales

and frequent turnover at vehicle dealerships often results in the further loss of vehicle salespeople along with their vehicle customer relationships and knowledge. In this competitive vehicle environment, machine learning enables platforms to assist the vehicle sales team by tracking the vehicle consumer behaviors of each vehicle customer, learning and memorizing their preferences and predicting their future vehicle purchase needs.

Finally, I recommend that for a vehicle dealerships marketing platform to make their customer engagement efficient and fully-functional, I should be able to: applying (AI) tools to track every vehicle customer behavior across the web, connecting to a society of data sources, CRM, DMS, third-party, web vehicle brands, social email, click etc., aggregating and accurately cross-reference data from a variety of sources, leveraging this data to drive insights on a mass scale, as well as on an individualized basis, driving actions and automatically direct customer engagement via multiple channels based on where each customer is in their individual lifecycle.

How can (AI) provide businesses with better-informed decisions

I shall explain how (AI) technology can provide businesses with better-informed decisions to drive top-line growth, deliver meaningful experience for customers and smooth their path along the consumer journey. The widely understood definition of (AI) involves the ability of machines or computers to learn human thinking, reasoning and decision-making abilities.

A Narrative science study in 2015 year identified that (AI) was being used primarily in voice recognition, machine learning virtual assistants and decision support. This study also highlighted the many branches of (AI) and that techniques and their definition are used interchangeably. It is possible that (AI) can be used to gather big data , then to analyze to help businesses to predict consumer behaviors. For example, one of the most common techniques is machine learning, where algorithms are used to perform tasks by learning from historical data. Another growth branch of (AI) is natural language procession.

However, during 2017 year, search engines will begin to factor additional behavioral data into prediction of customer behavioral results, such as the user's history of searches and locations and previously captures conservations. Artificial intelligence will use this information to power predictive search results, e.g. predictive future consumer's choice behavioral processing for any kinds of businesses.

Predictive search will improve the quality of search results, and provide new insights into consumers' behavior and the moments which matter to them. Search will give recommendation into tailored how consumer individual choice in consumption process. Several of the largest online platforms already use machine learning to improve predictive consumer behavioral search results.

For example, Google's rank brain technology adds research by understanding the context in which the consumer has entered it. Over time, rank brain will learn further from user behaviors Amazon's DSSTNE (pronouned destiny) learns from shoppers' purchasing habits and consumption behavior to offer better product recommend actions, which Amazon can offer before a consumer has entered anything into the search bar. However, this technology is not independent of human input. For example, Google engineers will periodically retain the rank brain system to improve the models it uses. For another example, in 2016 year , Apple computer revamped its photos app to allow consumers to search for specific items in the phots, they want to find, not just dates and locations. Each photo that an intelligent phone or intelligent pad user takes goes through 11 billion computations, so that photos can understand exactly what is the photography.

It seems that in future, (AI) machine learning will allow search to evolve even further. Search engineers will deliver refined recommendations to their business users and use less human input to predict consumers' needs. For IBM computer example, it indicated 90% of the data that exists today has been created in the last two years. This huge explosion of data gives brands the opportunity to quickly spot and react to the latest trends, fashion and fads among its clients and potential clients. This will allow companies to better engage with younger consumers, who gain influence access to the latest trends, and use the brands. They associate with to help define who they are as individuals. Thus, brands have to identify and make use of them before consumers move on, but the vast quantity of data available makes. This a resource-intensive task. For next example, Lesara, a based online clothes store, uses this machine learning to inform its product decision often gathering information from internal and external sources. When its trends -spotting shoes. Lesara has a range of over 20 styles and sells hundreds of pairs a day. It focus on giving consumers, the very latest trends allow Lesara to develop on average of 50,000 new items each year. It compared to 11,000 old items each year. Thus, (AI) brain seems to human brain to own analytical ability to predict consumer behaviors.

For another example, Lesara is one online clothes store, uses machine learning decisions after gathering information from internal and external sources. One of its most popular products, shoes with LED started life when its trend spotting software flagged up a blogger wearing similar shoes. Now Lesara has a range of over 20 styles and sells hundreds of pairs a day. Its focus on giving consumers the very latest trends allows Lesara to develop an average of 50,000 new items each year, compared to 11,000 for its competitor Lara. it seems (AI) machine learning can help Lesara business to predict what kinds of shoes design or style that shoe consumers will prefer choose to buy in future shoe market trend. Thus, Lesara can predict shoe consumers' taste successfully and it can manufacture many attractive style of shoes. (AI) machine learning can gather global past shoe consumer's shoe shopping experiences, then analyzes to make conclusion to give lesara recommendation successfully. This will make the experience more enjoyable for shoe consumers and allow Lesara to advert whose different new style or design of shoes to deliver them move relevant messages by understanding the context of the experience.

However, (AI) machine learning will have this risk who manufacturers need to concern if they applied this technology to predict consumer behavior. It is on sample consumers' privacy issue, in order to avoid complaint chance occurrence. However, machine learning can tie this data together to identify which f the billions of devices are being used by individual consumers. This helps brands understand how consumer engagement and actions can be attributed to different messages in different contexts and at different time. So, machine learning can help brands to build confidence to promote their products by any advertisement channels. When, this new (AI) machine learning technology can conclude how to design their products to be the most attractive, due to it has more accurate to predict consumer behaviors to compare human themselves prediction judgement effort. It seems that (AI) machine judgement effort is more accurate to compare to human judgment effort.

For example, google is moving away from cookies and using logged in data to track and make to users. It plans to expand the scope of the brand lift tool from online video. Thus, consumers are responded will to shippable context, finding it persuasive and easy to navigate by (AI) machine learning decision. For example, fashion brands can aggregate their You tub videos and blogs into a mobile context marketing experience, such as brand centric context into a personal shopping activity gives the shopper an experience, who are likely to remember and tell their friends about any new style of products design promotion from these internet advertisement channels after (AI) machine learning tools' styles of product design recommendation.

What is (AI) deep learning techniques to forecast environment behavioral consumption

The (AI) deep-learning technology leads to performance enhancement and generalization of artificial intelligent technology. It influences the global leader in the field of information technology has declared its intention to utilize the deep-learning technology to solve environmental problems, such as climate change. So, it will help agriculture farming businesses can raise any plant food: vegetable, fruit, rice which grow up very easily if farmers can apply (AI) deep-learning technology to solve environment problems to influence their plant food grow. If the whole year seasonal change is very good and it is suitable for any plant food to grow in farming land easily, e.g. rain is enough and soil is enough for any plant food to grow in the farm lands. Then, fruit, rice, vegetable etc. agriculture businesses will have much beneficial attribution to global farmers.

The question is how to use deep-learning technologies in the environmental field to predict the status of pro-environmental consumption. We predicted the pro-environmental consumption index based on Google search query data, using a recurrent neural network (RNN model). To certify the accuracy of the index, we compared the prediction accuracy of the RNN model with that of the ordinary least square and artificial necessary network models. For example, the RNN model predicts the pro-environmental consumption index better than any other model. we expect the RNN model to perform still better in a big data environment because the deep-learning technologies would be increasingly as the volume of data grows. So, deep-learning technologies could be useful in environmental forecasting to prevent damage caused by climate change to influence any rice, vegetable, tomato, potato, fruit etc. different plant food grow in any countries' farming land easily.

For South Korea example, over 800 government agencies spent 2.2 trillion Korea won on eco-products in 2014 year. However, green products are rarely purchased outside these agencies. This phenomenon occurs because there is a gap between consumer attitudes and behavior , that is environmental attitude is a major factor in decision making vis-a-vis the consumption of " green" food and services (Jorea Ministry of Environment, 2015). Therefore, it is necessary to understand those consumer attitude, that will lead to sustainability-conductive behavior and consumption.

Environmental consumption prediction

Recently, many researchers have studied pro-environmental consumption and household indexes as well as suicide rate predictions using messages posted by internet users on Google trend, Tweets etc. channel. Whether can environmental consumption be predicted by (AI) deep-learning technological internet channel? How can impact the pro-environmental consumption attitudes of green policies? Korea scientists estimated pro-environmental attitudes using search query data provided by Google trend and confirmed through regression analysis, that pro-environmental attitude has a positive correlation with the pro-environmental attitude index. They also explained that environment-friendly attitude of residents plan an important role in policy making. In the past, most household consumption indexed were calculated through surveys, but (AI) deep-learning technological tool " big data" have recently gained research attention (Lee et al. 2016).

It seems that (AI) deep-learning technology can help agricultural export countries‘ farmers , e.g. US, UK, Canada, New Zealand, Australia, Japan, China, India etc. they can predict environmental behavioral consumption to any rice, tomato, potato , fruit, vegetable etc. plant food consumers. The beneficial advantages to them include as below:

(a) Assuming they know their countries' weather, when it has less rain to cause drought or when it has more rain in any seasonal time in the year. They can choose not to grow any kinds of above these plant food to avoid loss.

(b) They can make any kinds of above these plant food price raising after their prediction of these bad seasonal time to cause their plant food shortage supply challenge. Because these plant food consumers‘ demand number is more, but the supply of these above plant food supply number is less. However, due to they had predicted when the bad seasonal time can not allow them to grow these above plant food before. So, they have enough time to grow many these above plant food number in predictive good seasonal time to prepare to supply to their plant food import countries' plant food consumers to eat. Thus, these predictive environmental consumption plant food export countries can raise their plant food price to sell to them. When, the other non-pre-predictive environmental consumption plant food export

countries can not supply any one of those plant food to them to eat, due to the bad climate to cause them can't grow any one of these plant food to export to sell.
Thus, (AI) deep-learning technology can be applied to predict how to raise the plant food supply number in order to raise price to the import plant food countries consumers to eat, due to they feel difficult to buy these plant food to eat in the bad climate seasonal time in whole year.

(c) (AI) deep-learning technology can help climate scientists to find what reasons cause their countries; rain sudden increases or cause their countries' rain sudden decreases. After its gathering data analysis, it can assist climate scientists to find solution methods to attempt to control the rain level can be right falling down level to let agricultural export farmers who can grow their plant food to sell to agricultural import countries in whole year.

(d) The agricultural export countries' farmers can apply (AI) deep-learning technology to help them to choose whether growing which kinds of plant food in that whether climate time to earn more plant food consumption number more easily.

Due to the agricultural countries climate will often change, for example, tomato, potato, rice, fruit etc. plant food can be adapt to grow in more rain time, but vegetable can not be adapt to grow in more rain time. If farmers can apply this technology to predict when it will have move rain or when it will have less rain to fall down in their countries. Then, they can choose to grow which kinds of plant food number more, in the suitable seasonal climate time in order to raise plant food growing number productivities to supply to sell to satisfy any agricultural food import countries' demand effectively.

(e) (AI) deep-learning technology can help agricultural import countries to solve agricultural food shortage challenge in long term. When this technology can be popular to base applied by the agricultural plant food export countries. It will solve global agricultural food shortage challenge. For example, when one agricultural export countries' farmers can popular accept to apply this technology to predict when to grow which kinds of plant food more to rise number productivities to sell. e.g. vegetable, fruit, rice Besides another agricultural export countries' farmers can also accept to apply this technology to predict when to grow plant food, e.g. potato, tomato to raise number productivities to sell. Then, they can concentrate on growing the specific kinds of plant food in order to raise the specific plant food number productivities in every seasonal change time every month. Then, global agricultural plant food supply must be raised, due to these predictive environmental change farmers can know who ought grow which kinds of plant food to sell to raise number productivities.

How can apply (AI) digital channel to predict consumer behaviors?

(AI) digital channel can be applied to help businesses to evaluate whether how much the product price is the most attractive to persuade consumers feel it is the most reasonable price to sell. It helps consumers to feel which brands of products which ought change the price to let consumers to choose to buy the brand of product. It can be applied to predict whether how many consumer numbers can be increased or decreased when the brand of product's price is variable. It aims to give opinions to help any brand of product manufacturers or sellers to judge whether which price is the most reasonable to let consumers to accept to choose to buy the brand of product in popular.

Thus, (AI) price measurement technology can be preference to be applied online communication ecommerce and mobile phone internet platform aspect. As businesses can enter their past products prices data and past customer number data into computer or mobile. Then, (AI) price measurement technology can gather these data to analyze these product prices and past customer number to compare their prices variable changing range level to find their price variable difference to measure to make conclusion about every product's price variable changing will influence how many customer number increase or decrease changing to choose to sell their different kinds of products more accurate. Then, (AI) price measurement software will help them to analyze all past price variable changing data to compare whether which price range can let customers to feel it is more reasonable and attractive to influence them to choose to buy the product among different brands of product choice.

Because any product's price is one important factor to influence consumers to choose to buy the product, instead of quality, durability, shape, appearance, color, brand familiarity etc. factors. Any online businesses with a focus on Asia should considerate (AI) customer care, and virtual shopping experience, whereas is Europe and North America still value face-to-face and/or real human interaction over (AI) or virtual worlds.

For example, Amazon publish has applied (AI) price measurement technology to help authors to decide how much every different topic of e-book or paper book price, it can attract the largest number of readers to buy. Any one author only needs to type whose book name to Amazon publish author himself/herself Amazon website. Amazon publish (AI) price measurement learning machine will help them to auto-calculate and judge how much e-book or paper book price is the most attractive and the most reasonable in order to increase reader number to buy their e-books or paper books to read. So, (AI) online price measurement machine will gather past similar book names and past every similar book readers' reading times and the number of readers to give opinions to let every author to judge whether his/her very new e-book or paper book ought charge how much price to the e-book or paper book which can attract many readers to choose to buy. Although, it is not ensure that the e-book or paper book price must let readers to feel it is the most reasonable price to choose to buy in reader's view point. However, it has other factors to influence readers' choice to buy the e-book or paper book, e.g. whether the book content is attractive to public, the author's familiarity, the book's page is enough or not to satisfy readers to read etc. factors. But, instead of all these extra factors to influence readers to choose to buy the book to read. (AI) price measurement learning machine can real give opinions to every author to let them to judge the e-book or paper book different price range whether is too high to influence readers to choose to buy to read or tool low to influence readers feel it is possible poor content book to compare other similar content books. Thus, (AI) price measurement machine can help authors to predict every reader's reading behaviors or reading experience and reading habit from online channel in short time easily. The author only enter the book name to let Amazon publish price measurement machine to check, it will follow past reader's reading habit and reading experience to judge whether the similar all book topic sale record to judge how much price is the reasonable price to attract many readers to buy the book.

Hence, (AI) can be applied to digital channel to help businesses to predict consumer behavior in the future. In the future, mobile/smartphone, laptop, desktop will be most frequent used ecommerce channels to develop online business. So, (AI) can be also applied to these platforms to gather data to make analysis to help businesses to predict consumer purchase behaviors popularly. Due to , ecommerce is popular to global, so digital online and instore channels can be one good channel to let (AI) learning machine to make platform to gather past every online consumer purchase (buying) experience data to help businesses to build brand personality and having a responsible, positive impact on society.

To apply (AI) learning machine technology to understand customer online purchase behavior, it will raise business e-commerce successful chance: For example, (AI) learning machine can help businesses to gather data to analyze to determine whether short-term or long-term signals in the online consumer behavior that indicate higher purchase intents to let every online business to know. (AI) learning machine can find that online users with long-term purchasing intent tend to save and click through on more content.

However, as online users approach the time of purchase their activity becomes more topically focused and actions shift from saves to searches from online consumption channel. Then, (AI) learning machine will further find that the brand product purchase signals in online behavior can exist weakness before an online purchase is made and can also be traced across different online purchase categories. Finally, (AI) learning machine synthesize these insights in predictive models of online user purchasing intent to the brand of product. Taken together, it's work identifies a set of general principles and signals that can be used to model online user purchasing intent across many online content discovery applications. Thus, (AI) learning machine can help online businesses to gather any online users' click online behaviors data to judge whether there are how many online users will choose to find their online business websites to make final decisions to buy their products from online channels. Then, it will give opinions to help the online businesses to let it to judge whether what are the important website factors will help its online business to attract many online consumers, e.g. designing unattractive website issue, online unattractive product photos issue, unclear website color issue, unclear website advertisement message, contents and words impressions issue, lacking image movement frequent attractive seeing issue etc. different website factors. Thus, online digital channel will be one good choice to apply (AI) learning machine to help businesses to predict consumer behaviors.

Can apply artificial intelligent learning machine " big data" gathering method to predict manufacturers' behavioral performance ?

In consumer view point, can they apply (AI) learning machine to predict manufacturers' behavioral performance to judge whether whose products are value to buy. Nowadays, (AI) and big data are reshaping the risk in consumer privacy. For example, consumers want to hide their willingness to pay just as firms want to hide their real marginal cost, and buyers have less favorable information, say a low credit shore, prefer to withhold it just as sellers want to conceal poor product quality. So, it implies that it is possible (AI) learning machine can help customers to gather any manufacturers' past sale performance, e.g. how many complaints or appreciation from clients, product quality etc. sale data to let consumers to make judgement whether it is value to buy to compare other competitors. So, it has risk to the poor product quality of manufacturers. Otherwise, it has benefits to the good product quality of manufacturers. It also implies all manufacturers' privacy is not protected or secret when (AI) learning machine is popular to be used to predict manufacturers' behaviors by consumers.

Information economists suggest that both buyers and sells have an incentive to hide or reveal private information, and these incentives are crucial for market efficiency. Data technology that reveals consumers type could facilitate a better match between product and consumer type, and data technology that helps buyers to assess product quality could encourage high quality production.

Thus, (AI) big data technology can also assist consumers to gather different manufacturers' data to compare what their advantages and disadvantages of their products are. Then, consumers can make comparison to choose which brand of product is the suitable to whom to buy in these more choice consumption market. (AI) learning machine will gather similar brand their products' data to analyze to make conclusion to let consumers know or feel to make final judge to find what advantages or disadvantages of these sample brands of similar products' comparison from internet. On the other hand, it means that manufacturers can gather consumers' past purchase behaviors or purchase experience from (AI) big data gathering method to record and analyze to give opinions to let manufacturers to know what reasons or factors influence consumers choose not to buy their products from internet.

(AI) big data gathering consumer behavior prediction method can give these benefits to manufacturers and consumers both, such as: New concerns arise because (AI) technological advance which have enables reducing cost of collecting, storing, processing and using data in mass quantities extend information beyond a single transaction. These advances are often summarized by the big data, it means charge volume of transaction-level data that could identify individual consumers by itself or in combination with the datasets.

The popular (AI) takes big data as in input in order to understand, predict and influence consumer behavior. Modern (AI) is used by legitimate companies, could improve management efficiency motivate innovations and better match demand and supply. But (AI) in the wrong hand, also allows the mass production of fraud and deception. Since , data can be stored, traded and used long after the transaction. Future data use is likely to grow with data processing technology, such as (AI) big data gathering consumer and manufacturer behavioral prediction method from internet channel.

Thus, future (AI) big data learning machine can also help consumers to choose the best brand of manufacturer's products among different brands of manufacturers products choice to compare their past sale performance from internet. They can apply (AI) big data statistic method to gather all different manufacturers' similar products past sale data to compare their advantages and disadvantages to make the best decision to choose to buy which brand of product is the most suitable to them to buy to use. It seems (AI) big data can also help consumers to predict any manufacturers' manufacturing behaviors or manufacturing performance whether they are improving their product quality or are deteriorating their product quality. Thus, (AI) big data tool is also important to help customers to predict future the different brands of manufacturer performance will have improvement in possible.

Thus, I believe that artificial intelligent "big data" gathering method can be suggested to be applied to attempt to predict consumer behavioral changes in global business environment, the reasons are as below:

On the consumer's beneficial hand, Consumers can apply this method to attempt to gather any global manufacturers data to be analyzed by this artificial intelligent learning system. Then, it analyzed all the different brands of specific similar product manufacturer' data to compare what are the range of the best past manufacturing history and sale data to the group of best manufacturers, and what are the range of the better past manufacturing history and sale data, and what are the range of the good past manufacturing history and sale data, and what are the range of the common past manufacturing history and sale data. Finally, the (AI) learning system will compare all the specific similar product, e.g. mobile phone or computer, television, car etc. different kinds of specific products of global manufacturers to conclude the result is such as whether which brands will be the best manufacturers to let the consumer to buy the television or mobile phone or computer or car etc. different kinds of products. It can make more accurate judgement to compare general human's phone or questionnaire surveys investigation method, newspapers, television, radios, internet searches etc. different manufacturing news or data gathering channels to find which brands are the most worth confidence to consumers to choose to buy the specific product in the global consumption market.

On the manufacturers' beneficial hand, manufacturers can apply (AI) data gathering method to predict consumer emotion and buying behavioral changes more accurate. For example, the vehicle manufacturer, it plans to gather data to predict potential driving fast speed sport vehicle consumers' preferences trends in order to make the accurate judgement how to design its sport vehicles to attract many sport vehicle buyers who will choose to buy it's brand of any driving fast speed sport vehicles. It can attempt to apply (AI) intelligent learning system to gather global different brands of sport vehicle data concerns that all past driving fast speed sport vehicle buyer's preference of sport vehicle design. Then, the (AI) intelligent learning system gather global different brands of driving fast speed sport vehicle which had ever been purchased by the different country's driving fast speed sport vehicles consumers. After, it can compare divide the range of similar driving fast speed sport vehicle design and similar price to be different groups. The (AI) intelligent learning system can attempt to follow the past number of different brands of driving fast speed sport vehicle buyers to calculate how many driving fast speed sport vehicle buyers who choose to buy the brand of driving fast speed sport vehicle as well as it will analyze and make judgement to find whether the cheaper price reason attracts the different countries sport vehicle buyers choose to buy the brand of driving fast speed sport vehicle or the attractive design reason attracts the different countries sport vehicle buyers choose to buy the brand of sport vehicle or fast speed reason attracts the sport vehicle buyers choose to buy the brand of sport vehicle.

For example, although some brands of driving fast speed sport vehicle manufacturers' prices are very high, but they can still attract global many sport vehicle consumers to buy. Whether all sport vehicle's attractive design is the main factor to influence them to buy or whether it's fast speed is the main factor to influence them to buy or whether it's safe confidence it the main factor to influence them to buy or it's familiarity brand is the main factor to influence them to buy. (AI) intelligent learning system will attempt to make judgement and analysis to conclude whether the attractive design factor is the main factor to influence many sport vehicle consumers to choose to buy the brand of sport vehicles.

Otherwise, for another example, although some brands of driving fast speed sport vehicle manufacturer's prices are low, but they can not still attract many global many sport vehicle consumers to buy. Whether all vehicle's unattractive design is the main factor to influence them choose not to buy their fast speed driving sport vehicles or whether the unsafe factor is the main factor to influence them choose not to buy their fast speed driving sport vehicles or whether unfamiliarity brand is the main factor to influence many consumers choose not to buy their fast speeding sport vehicles.

Thus, when (AI) learning system had helped the fast speed sport vehicles manufacturer to gather all different brands of fast speed driving sport vehicle's past sale data and price data, design of different sport vehicle, e.g. color choice, method of style, comfortable chair styles and chair sizes and what kinds of steel material to manufacture the sport vehicles data and driving safe and accident occurrence data and the data concerns what reasons of the past complaint to brand of sport vehicle manufacturer from its sport vehicle buyers. Then, it can make more conclusion to give more accurate opinions whether which brands of fast speed driving sport vehicle manufacturer(s) whose sport vehicle design is the main factor to attract consumers choose to buy its any driving fast speed sport vehicle products really. Thus, it seems that it can make more accurate judgement to compare television survey, questionnaire survey to gather data concerns how to design the fast speed sport vehicle to attract consumers to choose to buy the sport vehicle manufacturer's planning sport vehicle products. I believe that (AI) learning system can help the sport vehicle manufacturer to make more accurate conclusion or judgement how to design its fast speed driving sport vehicles to attract it's consumers more easily.

(AI) tool predicts consumer immediate and expected emotion how to influence consumption decision

If (AI) tool can be confirmed to apply to predict consumer behavior, then I can conclude that it can be attempted to apply to predict what the factor(s) of the product itself can cause the consumer has positive or negative emotion, so the manufacturer can attempt to avoid the bad factors cause to bring negative emotion to influence the consumer chooses not to buy the product more easily, such as vehicle product case.

Economists refer to the consumption desirability is as " utility" and the product or service consumption decision making is arose influenced by maximizing utility only. However, they neglect consumer individual immediate emotion change will also influence the consumer individual consumption decision consequently. Expected emotions are those that are anticipated to occur as a result of the outcomes associated with different possible courses of action. For example, if a potential investor, were deciding whether to purchase a stock, who might imagine the disappointment who would feel if who ought it and it reduced its price. Otherwise, whose emotion would experience , such as regret if it increased in price, but who does not buy it before the stock rise its price. However, I believe nowadays technology, in the future one day, (AI) tool can be attempted to assist consumer psychology profession or marketing research profession to assist them to find what are the bad factors to influence consumers choose not to buy any manufacturers' products. Then, when the manufacturer can discover what are the bad factor(S) cause(S) consumers who do not choose to buy their products, then the manufacturer can raise whose product of consumption desirability or " utility" to raise whose product's consumption decision making is influenced by maximizing utility. Hence, (AI) tool will be possible to find what the bad factor(S) to cause consumers do not choose to buy the manufacturer's product in order to raise the product's utility to bring consumer positive emotion to choose to buy its product in possible. SO, (AI) tool will be one consumer psychological emotion prediction tool to assist any manufacturers to help their products to build positive emotion to any consumers in possible.

The key feature of expected emotions is that they are experienced when the outcomes of a decision materialize, but not at the moment of choice, at the moment of choice, they are only feel about future emotion. Such as consumption case, if the consumer chose to buy the product or consume the service before it's price is increased. Then, the consumer will feel happy and it is worth to purchase or consumer the service as well as the consumer's expected emotion is positive before who decides to buy the product or consume the service, because who believes or feels the product or service's price will be raised in short term, e.g. after one month, one week. Thus, it means that if the consumer does not believe or feel or predict the product or service's price either it will increase or decrease in short term, whose emotion will be negative, those negative emotion will influence who does not decide to buy the product or service, it is possible that who feel it is not worth to buy the product or consume the service immediately. He She will choose to consume the service or buy the product to wait it's price is decreased later. it seems that the consumer's positive or negative emotion will influence who decides to buy the product or consume the service later or earlier. Thus, it has close relationship between the consumer individual immediate purchase or consumption decision and positive emotion or negative emotion (either expected emotion or immediate emotion influences).

Consequently, if (AI) tool can help any manufacturers
to predict when its product price ought to be increased or decreased in order to attract consumer to choose to buy its product. Then, it can help any manufacturers to build positive expected emotion to attract consumers to choose to buy its product more easily. For example, when the (AI) tool can predict when the consumer expects the product price will fall down, then it can give ideas to the manufacturer to raise up the product price in the month, then it predicts many consumers expect the product price will fall down after six months. So, the product price will not be fall down after six months. So, many consumers will feel disappointment and they will choose to buy the product if the manufacturer
decided to raise the product price after six months. Then, the higher product price will cause many consumers worry about the product price will continue rise up, so they will prefer to choose to buy the product immediately after six months because they afraid the product price will continue to rise up in the year. Then, I assume that (AI) tool has effort to predict when consumers feel the product will rise up or fall down, then it can give ideas to the manufacturer when to rise up or fall down the product price in order to attract or persuade many consumers choose to buy the product in different period in the year.

What does (AI) tool predict immediate emotion mean?

Psychologists indicate that immediate emotions, by contrast, are experienced at the moment of choice and fall into one of two categories. Integral emotion, like expected emotions, arise from thinking about the consequences of one's decision, but " integral emotion", unlike expected emotions are experienced at the moment of choice. Such as purchase stock case, the share buyer might experience immediate fear at the thought of the stock's losing value. " Incidental emotions" are also experienced at the moment of choice, such as a consumer predicts the product or service price whether it will be risen up or fallen down. If he/she feels the product or service price will fall down after next month and he/she will choose to buy the product or consume the service. But consequently, after next month, the product or service's price won't fall down absolutely.

Then, he/she will have incidental emotion to influence whom to choose whether he/she ought buy the product or consume the service, due to the product or service price is not still fall down. Otherwise, he/she is fear the product or service will not fall down in short term. Even, it will increase price later. Hence, whose incidental emotion will have possible to influence whom to choose to buy the product or consume the service after one month, if the product or service's price is still not increased absolutely. So, (AI) tool can be attempted to apply to predict when the product price ought need to be raised or fallen down in order to attract consumers to choose to buy the manufacturers‘ product in different period.

Economists indicate utility an individual consumption with an outcome might arise from a prediction of emotion: For example, a dinner eater might choose a higher utility to an Italian restaurant dinner than a French restaurant dinner because who anticipates being happier at the former, even the former's dinner price is higher than the French restaurant.

So, such as this restaurant dinner case, if one (AI) tool can assist the French restaurant owner to find what factor(S) cause(S) the dinner consumers do not choose to go to its restaurant to eat its food, e.g. high price factor, bad taste factor, bad wait service performance factor, bad cooker's cooking skill factor, poor advertisement promotion factor, poor familiar factor, poor location or poor eating environment etc. different factors. Then, the French restaurant owner can find methods to avoid the bad factor(S) cause(S) many dinner consumers do not choose to go to whose French restaurant to eat dinner more easily.

The question is that whether the positive emotion factor can influence the consumer changes whose mind to choose to consume the more expensive service or buy the more expensive product. To answer this question. it depends on whether the consumer has an imperfect understanding of whose own tastes or the consumer has a perfect understanding of whose own tastes to the product or the service.

It means the consumer will choose to buy the product or consume the service, even it's price is higher than other general similar products or services if who has a perfect understanding of whose own tastes to the product or service. Otherwise, who won't choose to buy the product or consume the service, due to it's price is higher than other general similar products or services if who has an imperfect understanding of whose own tastes to the product or service. So, it seems that the consumer's negative or positive emotion arise will be influenced by whose perfect or imperfect understanding of whose own tastes to the product or service factor.

It concludes that whether how much degree of the consumer's utility to the product or service. It is not the only one important factor to influence the consumer to choose to buy the product or consume the service. Otherwise, the consumer's imperfect or perfect understanding own tastes to the product or service factor will influence the consumer to arise positive or negative emotion to make final purchase or consumption decision immediately. It will be one more consumption influential factor to lead the consumer to make the final consumption decision making immediately. So, future (AI) tool ought to be innovate to own how to judge good taste or bad taste for any food in order to predict food consumers to choose to buy the food manufacturer's any foods more attractively.

(AI) tool technical innovation in cruise tourism
immediate positive emotion influence to
cruise travelling consumers

Can apply (AI) tool to cause positive emotion to cruise tourism consumers? Cruise tourism industry is the most influential emotion industry example to influence cruise travelling consumers' travelling entertainment choice. I shall indicate some evidences how it's innovation will influence cruise travelling consumers' emotion to be changed to positive from negative immediately as well as to prove how the cruise traveler higher utility feeling to the cruise tourism provider is not the main factor to influence whom to choose the cruise provider to consume whose cruise journey service arrangement.

Nowadays, cruising has become one of the fastest growing sectors within tourism, cruise service providers need have themselves unique different entertainment service arrangement to satisfy every cruise travelling consumer individual needs in order to attract every one to choose whose cruise arrangement easily, e.g. meals, activities, entertainment and varied destinations create one-stop holiday shop, reasonable competitive ticket fare. Hence, it seems it is one exciting emotion industry. If the cruise service provider can bring positive emotion to influence many cruise travelling consumers immediately. The, even it change higher service fare to compare other similar cruise service providers. I believe it won't influence them to choose other similar cruise service providers if it can often bring immediate positive emotion to its cruise clients during they are staying in its cruises or during they have left its cruises, but they will often remember or won't forget to enjoy their cruise service provider's happing time forever. Hence, if (AI) tool can be attempted to help cruise entertainment providers to arrange different cruise journeys for varied destinations , to arrange different entertainment facilities, to arrange the different taste food to satisfy different countries age cruise consumers' needs. Then, the (AI) tool will assist the cruise providers to bring positive emotion to let every different countries age cruise consumers to feel satisfactory in order to choose to the cruise providers' cruise entertainment service more attractively.

How can apply (AI) tool to predict cruise service providers bring positive emotion to their clients?

Future, (AI) tool can help any cruise providers to design these kinds of any one entertainment service arrangement to satisfy the cruise provider's customers' needs.

There are different special interests cruising , such as wellness at sea, freighter cruises, river cruises. It has increased the attractiveness of cruising: Romance is for lover cruise traveler target, luxury is for rich cruise traveler target, exotica is for enjoyment exciting feeling traveler target. So, every kind of cruise traveler target will have different kind of cruise entertainment service to satisfy their needs. If the cruise service provider can provide the right and attractive cruise entertainment service to satisfy the specific cruise target. Then, it will bring the positive emotion to the specific cruise target consumers more easily.

Cruise travel was shaped for mass tourism. Prices have been very differently segmented. There are basically four types of markets (Biederman, 2008):

- Contemporary market: On board fun and amenities are playing important role and destinations have secondary importance.
- Premium market: This category is more expensive than the contemporary category and where the destination has same importance as on board amenities.
- Luxury market: It was once dominant type of cruise tourism, but now it

has only a small portion of the industry. Generally, it is the most expensive cruise category and usually it takes longer than average cruise days.

- Adventure/exploration: It refers relatively long cruises with special and exotic places where the destination is the main purpose of the trip.
- European cruise travel: Duration takes more five days than worth American travel duration. There is a tendency on European market during the years that duration of travel is getting shorter. This short demand of is explained with the strong demand of customers (Hensen, 2003). Beside this, it is most likely that cruise companies try to convince tourists with short haul travels instead of long term cruise trips for more expenditure.

Thus, I believe that even, the cruise service provider charges higher ticket which won't influence cruise consumers who do not choose its entertainment service on its cruises. If it can arrange the attractive cruise entertainment facilities and destination journey arrangement, staying days arrangement to satisfy different specific cruise target market needs absolutely in order to bring whose emotion to be positive to it's service provision. Then, the cruise service provider will attract many potential cruise clients to choose its cruise service absolutely. Otherwise, if it only bring negative emotion to its cruise clients, it will not attract many potential cruise clients to choose it or loses its old cruise clients, even, its cruise ticket price is needed to decreased in order to raise competitive effort.

In conclusion, I believe that future (AI) tools need to learn how to bring cruise consumers to arise individual immediate or expected positive emotion, this positive emotion consideration is more important to compare to how to reduce cruise ticket price in order to attract cruise clients in global cruise competitive cruise industry.

Differentiation through the characteristics of cruising route method from (AI) tool route judgement

Future, (AI) tool can attempt to help any cruise entertainment service providers to judge how to design different route to attract different countries age cruise clients' choices to satisfy their cruise journey entertainment needs. The determinants of the cruising route's characteristics (functional, social, and emotion) is important factor to influence the cruise service provider's success. Cruising product is no longer selected primarily for the cruising service, but for the content of cruising route. So, the cruising route will influence the cruise consumer individual emotion, because it is the main service need for every cruise consumer.

The approach called the " land sea cruising in product development" is increasingly becoming an area of interest, e.g. determining the direction of the effects of the individual cruising route characteristics on service value's perception , and providing an evaluation model of the route's perception , and indicating significance variables of attraction.

The questions that cruise planners need to know: How does each of the identified determinants affect the overall perceived value of the cruise route? How the overall perceived value of the cruise route affects customer behavior intentions?

Because different routes factor will influence cruise consumer individual emotion changing seriously. It means the ship has become only a tool, when the offered route whose attractiveness highly influences the impression of the guests has become crucial.

Consumer behavior in cruising segment includes all the activities and influences in the selection of the specific cruise route. There activities result in decisions and actions related to a defined price, selection and reselection of cruising company (Cannot, Brink and Brijball, 2006).

How to apply (AI) tool to arrange cruise route planning have close relationship to influence cruise consumer emotion?

Firstly, use value of cruising routes is based on the subjective experience, and shows how individuals assess the route during, or immediately after sailing. It is affiliated with the benefits that cruising guest realize by choosing a route , and it is subjective because it depends on the individual assessment (photo taken on the route for one guest presents just a family souvenir, and for professional photographers are embodied financial capital).

Secondly, the utilitarian value is also subjective-oriented and is tied on the point where the inner and us ability of cruising routes are compared with the sacrifice of the client (money and time). Finally, the value is considered as the outcome of the comparison of scarifies and personal benefits, which is resulted in essentially utilitarian nature.

Hence, route design is the main value of cruising tourism and it is primarily determined and analyzed from the aspect of observed customers. Otherwise, the cruise is only one tool to be caught for the cruise passengers, whether the cruise can let whom to sleep comfortable , providing what kind of food to them to eat, what kind of entertainment facilities are provided to them to play, these issues are not more important to compare how to design route to bring them to travel to anywhere to enjoy in this cruise journey factor. Because how to design the route factor can bring each cruise passenger to influence them to feel either negative or positive emotion directly. The whole route journey planning is the most influential factor to influence the cruise passengers to feel whether they ought choose it's service again or not in the future.

(AI) tool judges the difference between utility factor and emotion to influence consumer decision making

In economic utility or immediate (expected) emotion aspects, whether which is more influential to excite consumption. To analyze whether it is economic utility or immediate (expected) emotion more influential to excite consumption. It depends on the consumer individual consumption choice is in which situations. For example, if the industry's general consumer individual consumption decision is concentrate on emotion influential aspect, such as cruise entertainment industry, hospital care service industry, theme park entertainment industry, movie watching entertainment industry etc. Above all these industries have same nature, it is service. So, it seems that service industry's main influential factor is immediate (expected) emotion influence, it is not economic utility influence. Otherwise, product sale industry's main influential factor is utility.

(AI) judges consumer utility factor

For this toy choice situation example, parent choose to buy one toy to give whose child to play. They usually considerate which kind of toy is attractive to their child whom like to play. In many different kinds of toys choice, if the child likes to choose the kind of toy to play. After the child's parents had purchased the kind of toy to let whose child to play one period time, e.g. six month. Then, when the child feel that who has need to buy another new toy to play, due to he/she feels bored to play this toy. So, it seems that the child feels this toy has less utility or it's utility is decreased. So, he/she expects whose parent can buy another new kind of toy to let whom to play. It also implies that it is not emotion factor to influence the child to feel boredom and unfunny to play this kind of old toy after six months. It is the product's utility factor which can not attract the child to play it any more. So, this old toy's utility is decreased when this child spends six months to play it. This toy's value is only six month utility to this child to play. Otherwise, if this kind of toy is bought by another parent. It is possible that the another child like to play this kind of toy one year or more. So, it's utility to another child is one year or more period. So, product's utility period is difference, it depends on how long time of the user's satisfactory time.

As this toy case, the child's decision will influence whose parent choose which kind of toy to buy to whom to play. Usually toy price is not difference too much. Parent won't consider when the toy price will be increase or will be decreased to influence their emotion to decide not to buy the toy immediately. So, when the child like to play the kind of toy, even the product's price is more than other kind of toys, and the parent feel it is possible that the kind of toy's price will be fallen down later. They will still choose to buy the kind of toy to let their child to play, they won't be influenced not to buy this kind product by later cheap price factor. So, immediate emotion is not the main factor to influence this parent does not choose buy this toy at this moment. Otherwise, utility factor will influence when the parent will buy another new kind of toy to provide to this child to play. If the child enjoy to play it only three months, after he/she will feel bore and he/she will tell whose parent to buy another new kind of toy to let whom play when the fourth month is beginning. So, it implies that if the kind of toy product can have more attractive utility time, then it can attract many parent to choose to buy it among different kind of toys. Thus, when this kind of toy's utility time is longer time. Then, it is possible that it can influence many parents choose to buy it's different style or design of similar kind of toys to let their children to play. In general, when many parents accept to buy this kind of different style or design of similar toys to give their children to play. Due to it's popular long time utility factor, it will influence children like to play it longer time to compare other kind of toys. Consequently, it will influence parents do not need often spend too much money to buy other kinds of toys to give their children to play. So, longer time utility factor to the product can attract many consumers to choose to buy the kind of product to compare lesser time utility factor to the product. Hence, it proves the explanation why utility factor is the main influential factor to influence the consumer choose to buy the product.

(AI) judgement tool of Medical care and utility case

Medical care is an input in producing health, it is subject to law of diminishing marginal productivity. Health yields utility to the consumer. It is subject to law of diminishing marginal utility. It bring this question: Does either the patient's emotion or the medical care service or medical care product utility which one can influence the patient's hospital choice more?

To answer this question: We need to know medical care is one kind of nursing care service in hospitals or clinics and medical care product is one kind of medical care product sale from merchants, e.g. medicine or medical equipment. So, in medical industry which has different kind of medical care services to provide to patients in hospitals or clinics as well as which has different kind of medical care products sale, e.g. medicine or wheelchairs, heart health measurement equipment etc. different medical care products in medical health industry.

In medical care aspect, it is one kind of any medical care service to patients from hospitals or clinics. So, medical care is an input in producing health service to patients from hospitals or clinics. When the patient is admitting to hospital or clinic, who needs to see doctor and the doctor need to give the right medicine to the patient to eat to ill whose illness. Even, if the doctor feels the patient whom needs to live hospital for one time period. Then, the hospital nurses must need to take care the patient during he/she is living in the hospital period. Consequently, if the patent can be health in short time, e.g. within one week leaving time, then he/she will be shortened time to leave the hospital in next weak. Otherwise, if the patent can not be health in short time, e.g. within on week, then he/she needs to live the hospital more than one week, even, one month, three months or more. So, the staying hospital time will influence the patent's emotion to feel whether the doctor's effort. If he/she needs to live the hospital long time, he/she will bring negative emotion to feel the doctor's medical effort is not good. The doctor's medical effort can not achieve or satisfy whose expected emotion during whose staying hospital time.

Thus, medical care is an input service in producing health, it is subject to law of diminishing marginal productivity. When the patient does not need to live the hospital longer time, the patient will feel more satisfactory to the hospital's doctor and nurses' care effort as well as the patient can give less money to spend the expenditure to live the hospital. So, the law of diminishing marginal health productivity will explain the hospital will shorten time to the staying days of the hospital to the patent as well as the patient's care expenditure will be decreased when he/she only needs to live to the hospital in short time. Otherwise, the patient needs to live longer time in the hospital, it means that the diminishing marginal health productivity to the hospital, the patient's staying hospital days will be increased and the patient's medical expenditure to the hospital will also be increased. It will bring negative emotion to patient and why this negative emotion factor will influence the patient would choose another hospital to live or find other doctors to see if he/she felt illness in future one day.

In medical care product aspect, health yields utility to the consumer. It is subject to law of diminishing marginal utility. Because patient needs to buy different kind of medical equipment to use or medicine to eat to attempt to cure whose illnesses. So, if the patent can choose the right medicine to eat from the doctor's recommendation or if the patent can choose the right medical equipment to use from the doctor's recommendation. When the patient buy less number of medicine to eat, then he/she can be health or when he/she buy the medical equipment to use, then he/she can be health. Then, he/she will spend less money to buy medicine to eat or medical equipment to use and he/she can be health in short time. Then, the medical consumer will feel the medicine or medical equipment has good utility to satisfy whose medical needs. So, good medicine and good medical equipment can only need short time and less money to let the medical patient to be health.

Immediate (expected) emotion factor

As cruise entertainment case, every cruise journey must provide fixed stay days on the cruise to let every cruise passenger to play to every cruise journey. So, cruise passenger can not change or extend whose fixed stay day choice in every cruise journey, when they had caught the cruise to go to sea on the day. Is implies that cruise entertainment has none longer time utility factor which can influence each cruise passenger's choice to each different design of cruise journey arrangement.

If the cruise passenger feels very satisfactory and enjoyable to the last time of specific cruise journey arrangement, e.g. five days and four nights New Zealand and Australia cruise journey. Due to this cruise journey can bring positive emotion to let the cruise passenger to let whom feel that he/she can not forget or remember this happy cruise journey forever.

It is possible that this time happy five days and four nights New Zealand and Australia cruise journey will bring positive emotion to influence this cruise passenger to choose to find this cruise service provider to help whom to arrange this same cruise journey or another similar cruise journey again after one month, or three month, or six month or one year or more. Due to this cruse passenger felt this cruise service providers' cruise journey design arrangement can satisfy whose needs and it can achieve whose expected emotion to be positive. So, this cruise entertainment industry must be immediate (expected) emotion influential factor more than time utility factor to influence the cruise consumer's cruise service provider and cruise journey choices.

In conclusion, it is not only utility factor can influence consumption decision. It is emotion factor can also influence consumption decision. It depends on situations whether the consumer is choosing to buy one product or consume one service. If the consumer is choosing to buy one product, how long time of the product's utility factor which will influence the consumer choose to buy which product. If the consumer feels the product can give longer utility time among other similar products, then he/she will have more chance to choose to buy the product. Otherwise, if the consumer feels the product can give lesser utility time among other similar products, then he/she will have less chance to choose to buy the product. If the consumer is choosing to consume one service, emotion factor will influence the consumer choose to find which service provider to consume the same service or similar service. If the service provider can provide excellent service to the consumer, then it will bring positive immediate or expected emotion to whom and it can attract the consumer to choose the service provider again. Otherwise, if the service provider can not provide excellent service to the consumer, then will bring negative immediate or expected emotion to whom and it can not attract the consumer to choose the service provider again.

Why and how (AI) judgement tool can judgement what utility factors are to influence consumer emotion

In emotion and utility both aspects, they include these situations. I shall explain how and why emotion and utility factors can influence consumption behaviors in these different situations as below:

(1) In the first situation is brand factor, the brand image, product quality, product knowledge , attitude and

(2) brand loyalty intangible factor will attract the consumer individual purchase.

For example, luxury

products, e.g. luxury fashion brands of clothing. Brands like Zara from Spain and H&M from Seweden began to produce catwalk-style fashion at low cost offering consumers of luxury fashion alternatives at low prices.

Nowadays, the luxury fashion sector is the fourth largest revenue generator in France, and one of the most remarkable sectors in Italy, Spain

, the USA and the potential markets of China, Russia and India. The luxury industry has increased having a huge youth in demand. The luxury consumer have much choice in products, shopping channels and pricing of luxury products. It has possible relationship of age, gender, income and other demographic factors with purchasing intentions to influence the rational and emotional buying behavior regarding luxury fashion products.

(2) The second situation concerns the decision-making of make or female consumers are possible experience an emotional desires and cognitive (reasoning) mind in purchasing choice process. Their emotion includes negative or positive buying emotion and mood management and cognitive process components include cognitive deliberation, planning buying with the exception regard for the future.

University had been using analysis of variances tests, male and female students were found significantly different with respect
affective process components including positive buying emotion, and mood management and cognitive process components include planning buying.

Significant differences were also found between the following product categories: shirts/sweaters, skirts, coats, underwear, accessories, shoes, electronic hardware, computer software, music , CD or DVDs, sports, memorabilia, health /beauty products and magazines/books for pleasure reading. No differences were found in regard to suits/business wear and entertainments.

The investigate proved that some products will have different emotion influence to cause female or male students whose final consumption decision to buy the kind of product. So, the difference od male and female students will have emotion influence to make purchase decision to buy the product in consuming choice process.

(3) The third situation concerns search advertising factor, e.g. online search to influence consumption behavior. Advertising is possible one method to persuade the consumer to choose to buy the product, even the consumer does not know the product exists. For example, proper cloth, a company based in New York, has a site on the social networking site Facebook.

Whenever the company posts a new photos of its clothes, all its face book " fans" automatically receive the information on their own face book pages.

"We want to hear what our customers have to say." It seems online advertisement is a potential promotion method to promote any new attractive products to sell to let publicity to know to buy. Internet is one popular communication tool to be used by youth today. So, when one company can have one website to let any youth to find and enter to the website to discover any new things easily. It will cause many consumption chances to let online potential clients to attempt to choose any products to make purchase decision from online advertising tool easily.

How does the role of advertising influence the purchase decision process? Needs and motivations are the starting points of purchase decisions. In fact, advertising is a communication of photo image, sound image, and word advertisement image channel to persuade consumers to choose to buy the brand of product or consume the brand of service between the merchant and its consumers from television, radio, newspapers, magazine, movie etc. channels.

Why does advertisement influence consumer
choice? For this case example, when one buyer waits until more information is gathered before making a decision. The time, two types of cost are involved. First, there are psychological opportunity costs experienced by consumers who are deprived of the product who need and are consequently in a state of psychological tension.

As time elapses, this psychological tension becomes more frustration. Second, buyers experience costs with the information-gathering efforts. They must invest time and energy to visit several retailers, seek out and read advertisements, or inquire for other opinions about the best product to buy.

These delayed decision costs considerably increase as time elapses. The buyer must seek information until it is felt that a search for additional information will bring about more costs than benefits. So, an advertisement is reaching a potential buyer when who is seeking information will have a greater impact, since the buyer is spending time and effort needed to seek out this information himself and he is less likely to find other competing and advertisements to obtain the additional information.

In general, buyers are generally more responsive to different brand advertisements, when they are seeking information on these brands. This is why the becomes a choice target for the advertiser provided the advertiser can identify and locate them. Thus, a client has interested and is in an information-gathering stage is asked.

Then, the advertiser takes advantage of the consumer's having identified him or herself to send a series of informative and persuasive messages or to send a salesperson who will try to conclude a sale. Thus, advertisement gives a chance to let consumers to gather information to choose the best product to buy or the most excellent service to consume.

What of situations do merchants need advertisements promotion? The short purchase cycle markets are characterized by routine purchase decision processes or by limited problem solving when a new brand is introduced on the market, e.g. coffee, bread , sugar, soft drinks, canned vegetables and household and beauty care products fall into this category. Another irregular purchase cycle markets are characterized by products that are purchased more or less regularly , e.g. cookies, cake mixes, wines, food products. Finally, long or unpredictable purchase cycle markets include all durable products, such as cars, household appliances and furniture (products from which occasions of purchase can't be predicted, which most consumers buy only occasionally). Hence, these kinds of products ought need advertisement promotion specially, due to advertisement can build brand image to let consumers to know. Especially , it is a new brand of product. In conclusion, advertisement will be one good channel to let new product to introduce its brand to let clients to remember in minds heart.

Consequently, future (AI) tool needs to learn how to design different kinds of advertisement to attract consumer attention to the product, needs to learn how to find what the bad factor(S) which cause(S) many consumers do not choose to buy the product, needs to learn when the product price needs to be raised up or fallen down in order to bring consumers' positive emotion to choose to buy the product immediately. SO, if (AI) tool can be invented to own itself effort to design different kinds of methods to predict consumer behavior in order to bring their positive emotion to the product successfully, then the manufacturer can earn positive consumer emotion advantage from the (AI) tool assistance for long term benefit to its product.

(AI) digital data gather technology predicts food consumer behavior's main barriers

What are the main barriers to food industry? When the food manufacturer applies (AI) big data gather technology to predict food consumer behavior? The barriers include that the food manufacturer / provider needs to decide whether when the right time is applied to the right (AI) digital big data prediction tool channel to find the right food consumers to be chose to full food consumption satisfactory questionnaires, how to gather multi-class food consumption classifiers on real-world food consumers transactional data from the food sale domain consistently to show the critical numbers of different kinds of food items at which the predictive performance most accurate? So, any food manufacturer / provider's advanced in (AI) digital data gather warehousing and management technologies can provide that opportunities for food business to enhance long term relationship with the food providers' clients.

However, food industry's (AI) digital data gather aims to improve food customer product targeting, increase food customer loyalty and food purchase probability to the food supplier. To effective identify, understand and satisfy the needs of their food customers, the food suppliers need to develop the right (AI) digital questionnaire questions and find the right food customers to fill every right questions from every digital questionnaire at the right time through the right channel.

Above of all these, they will be the barriers when one food supplier expects its (AI) digital data gather questionnaires which can conclude the most accurate prediction concerns any kinds of consumer food product choices. So, such as (AI) digital data prediction model, it is needed to incorporate into the food market segmentation, food customer targeting, and food challenging decisions with the goal of maximizing the total food customer lifetime. For example, (AI) big data gather transaction data is reasonable and accurate for building predictive models. Transaction data can be electronically collected and readily made available for data mining in lot quantity at minimum extra costs.

Suggestion to apply (AI) prototypes of food customer profiles method to predict food customer behavioral changes. Prototypes of food customer profiles mean to be extracted from the discovered bins and multi-class classifies models are built using those prototypes. The learned models can than be used to predict the class of food customer profiles (e.g. restaurants, school canteens, supermarkets etc. food suppliers) based on their food purchases. The approach is validated on the case study of a food retail and food service company operating in food and beverages market.

So, a food customer profile, it is a description (AI) data gather tool will record every of food customer using available information, which help in understanding their background and food consumption behavior. (AI) data gather tool can well develop every food customer profile, every food customer data is essential in food market analysis as they aid food suppliers in saving time and money by highlighting the real potential food consumers whose needs are to be met rather a range of individuals.

So, (AI) data gather tool can record every food consumer profile and every can be factual or behavioral food consumption. A factual food customer profile consists of a set of characteristics for (AI) big data gather record, e.g. demographic information , such as food customer name, gender, birth date, when a behavioral food customer profile consists of what the food customer is actually doing and is usually derived from (AI) digital transactional data gather record.

So, (AI) big data gather record's every behavioral food consumer profile can be much stronger predictor of the future food supplier consumption choice actions of a food customer. Furthermore, the food supplier's (AI) all past food consumer information that make up demographically based all past food customer profiles are expensive to acquire when the information for the food suppliers' past every food consumer food consumption behaviors. Moreover, food customer profile can be recorded to make real food purchase every time. So, when the food supplier finds the past food consumer's record from (AI) big data gather tool. Then, it can make more accurate judgement whether past every food consumer has chose to buy its food to eat how many times every year in order to predict whether its every past food consumer will choose to buy its foods how many times next year in possible. If the next year, its every past food consumer's consumption time to the food supplier is less than its current year consumption time. Then, the food supplier can attempt to find whether what factors to cause the past food consumers do not choose to increase food purchase times to the food supplier in current year. The factors may be possible be the food supplier's food prices are raised, food quality or taste is poor, the different kinds of food supply is shortage challenge, the food supplier's consumers lose confidence to buy the food supplier's foods to eat, when (AI) big data gather tool can help the food suppliers to find what the main factors to cause the past food consumer number to be reduced in order to predict how future food consumers' behavioral changes will be influenced from the food supplier's competitors in the global food supply market. Hence, (AI) big data gather tool can help every food supplier to attempt to find what the main factors to case the food supplier's food consumer number to be reduced as well as it can help the food supplier to predict how the food supplier's potential (past not every purchase its any food consumers) food consumers who can be persuaded to choose to buy its foods to eat by learning what the main factors influence.

In conclusion, (AI) big data gather tool can help the food supplier to find what the main factors influence its past food consumers do not choose to buy its food more times or find what the main factors will attract its potential (not ever buying its foods consumes) food consumers to choose to buy the food supplier's foods to eat.

The challenges of (AI) big data gather shaping the future of retail for consumer industries

Another challenge of (AI) big data gather is that how to shape the consumer behavior to let business owner to feel or know or predict. It means that how it express it's conclusion or opinion for every consumer behavior after it had gather all big data in any data gather period, e.g. three months, half year or one year consumer shopping model data gather period.

Because every kind of industry, consumers will continue to demand price and quality change , with a wide range of convenient fulfilment options among of different kinds of products or services supply. Overall, the (AI) big data gather procedure gives opinion concerns every time retail experience will become more exciting, simple and convenient, depending on the consumer's ever-changing needs. So, I believe that (AI) big data gather every conclusion or result will be different, due to consumer's price and quality demand will often change to every kind of product or service supply in retail industry. So, how to shape (AI) big data gathering's analytical conclusion or result more clear. I shall recommend organizations need to build great understanding of and a stronger connection to increasingly empowered consumers before they plan and implement how to apply (AI) big data gather tool to predict consumer behavior as below:

Firstly, (AI) is empowered by technology, the consumer is redefining value. The traditional measures of cost, choice and convenience are still relevant, but not control and experience are also important. Globally, consumers have access to more than 2 billion different products choice by a wide range of traditional competitors and dynamic new entrants, all experimenting with new business models and methods of client engagement.

As choice increases, loyalty becomes more difficult familiarity and the consumer becomes more empowered. Businesses will have no choice and constantly innovate and disrupt themselves by meeting new technologies of high standards and expectations of consumers. So, (AI) data gather tool will need to follow different target group of consumers' needs to follow their different kinds of product design or style choice preferable to gather data in order to conclude the different target groups of consumer behavior to give opinion more clear and accurate to let businessmen to understand more clear how its customers' behavioral choice trend in the future half month, even to two years period.

Secondly, businessmen need to adopt changing technologies rapidly. Technology will be the key driver of this retail industry. Industry participants will only success if they have a clear prediction to focus on how to using technology to increase the value added to consumers. They must , however, do so will I realistic assessment of their costs and benefits. Hence, (AI) big data gather technological tools will need to design to help them to gather data efficiently by these ways, such as the internet of things (IOT), artificial intelligence (AI) machine learning, augmented reality (AR)/virtual reality (VR), digital traceability. So, future (AI) big data gather tool are predicted to be most influential customer behavioral positive emotion changing tool for retail , due to their widespread applications , ability to drive efficiencies and impact on labor in order to impact consumer behavior changing effort from negative emotion to positive.

Thirdly, (AI) big data gather tool is an advanced data science of consumer behavior predictive tool. Businesses will have to bring the journey from simply collecting consumer data to using it to scale and systematize enhanced decision making across the entire value chain. When focused on their business goals, industry players should not lose sight of the impact that future capabilities and transformative business models may have on society.

However, (AI) big data gather tool will encounter these challenges when any business plans and implements to apply it to predict consumer behavior in retail industry. The challenges include that as below:

1. The high cost and difficulty of implementing new technologies . The (AI) big data gather tool needs capital and capabilities to be designed to implement to be applied to different retail industry users. so, expensive barriers to innovation, an organization and the skillsets of its people to support a new design of (AI) big data gather tool, highly digital technology may be required.

2. Slow pace of cultural change. Consumers need to adapt or accept (AI) new technology consumption model in the traditional retail industry. The rate of change is outpacing the ability of businesses to keep up. (AI) big

data gather tool needs to be designed to adopt in new or evolved business model requires, in most cases, a new level of customer behavioral predictive machine operation will impact to influence any retail businesses' consumer behavioral changes at a minimum, an organization's structure, capabilities, culture and decision making. If the retail business expects to apply (AI) big data gather tool to predict how to change its consumer behaviors and how their consumption behaviors will tend to change in order to achieve to change their positive emotion from negative emotion before they choose to buy its product or consume its service in success.

Challenge to using (AI) neural networks to predict customer behavior from big data gather tool

(AI) big data gather tool will encounter the challenge: How can predict customer behavior be represented as sequential data describing the interactions of the customer with a company or an (AI) data gather system through the time, e.g. these interactions are items that the customer purchase or views ? So, every customer data gather , (AI) needs to spend time to analyze how and why to cause whose consumption behavioral choice. It is too difficult matter or judgement for (AI) learning. So, (AI) needs to spend time to learn how to analyze every customer's shopping behavior or actin in order to gather all different consumers' past shopping action information in order to help business owners to predict future its potential customer shopping behavior how to change more clear and accurate prediction.

(AI) big data gather tool needs to learn to know that how to judge every customer interaction likes purchases over time can be represented with sequential data. Sequential data has the main property that the order of the information is important. Many (AI) machine learning models are not suited for sequential data, as they consider each input sample independent from previous ones. Therefore, at the end of the sequence, (AI) big data gather learn machines need to keep in their internal state of every customer purchase data, kind of product or service, price , whole year consumption times form all previous inputs, making them suitable for this type of data.

However, consumer behavior can be represented as sequential data describing the interactions through the time. Examples of these interactions are the items that the user purchases or views. Therefore, the history of interactions can be modeled as sequential data, which has the particular trial that an incorporate a temporal aspect. For example, if a user buys a new mobile phone, who might purchase accessories for this mobile phone in the near future or it the user buys a electronic book or paper book , he might be interested in books by the same author. Therefore, to make accurate predictions is important to model this temporal aspect correctly. To solve this predictive challenge of consumers to buy the product. One count the number of purchased products of a particular category in the last N days, or the number of days since the last purchase.

So, the (AI) big data gather designers can attempt to produce a feature vector which can be fed into a machine learning algorithm such as " logistic regression" will be the main feature and function to any (AI) big data gather machine to learn how to apply this " logistic regression" function or feature to predict any customer behavioral change for any product purchase or service consumption to the (AI) predictive consumer behavioral business users. Every different kinds of product purchases or services consumption will be needed to design " different model of logistic regression" in order to follow the kind of business to predict whose consumer purchase or service consumption behavior to predict more accurate.

Challenges of artificial intelligence, algorithms technology and machine learning impact to consumption market

Markets have played a key role in providing individuals and businesses with the opportunity to gain from trade. If (AI) big data gather tool can predict how to change potential customer behavior in success. The challenges to consumers will face that the overall market consumption model will be dominated by the businessmen only. So, it is not fair or reasonable to consumers, because (AI) big data gather tool has controlled or dominated all consumers' minds and it has predicted how and why every kind of product or service consumer shopping model or consumption behaviors how will change.

It will bring this questions: How can market designers learn the characteristics necessary to set optimal, or at least better, reserve prices after they had gather all data to conclude the analytical results of their consumers behaviors how will change? How can market designers better learn the environments of their markets?

In response to these challenges, artificial intelligence (AI) and machine learning are important tools for market design. For example, retailers and marketplaces , such as eBay, Amazon and many others are mining their vast amounts of data to identity patterns that help them create better shopping experiences for their clients and increase the efficiency of their markets. By having better prediction tools, these and their companies can predict and better manage dynamic consumption market environments. The improved forecasting that (AI) and machine learning algorithms provide help marketplaces and retailers better anticipate consumer demand and producer supply as well as help target products and activities for segmented markets. Another important application of (AI) 's strength in improving forecasting to help markets operate more efficiently is in electricity market example. To operate efficiently, electricity marker makers can attempt to apply (AI) machine learning tool to follow every household family electricity consumers' past electricity consumption record to judge (predict) how it will be every family's forecasting in the year.

An inaccurate forecast in the electricity supply and demand that can dramatically affect electricity market bad supply outcomes causing high variance in electricity charge prices or worse, blackouts. By better predicting every family's electricity demand and supply , electricity market makers can better allocate power generation to the most efficient power sources and maintain a more reasonable electricity stable charge market. Any example is design market, the application of (AI) algorithms to market design are already widespread and diverse.

(AI) algorithms technology , it is a safe that (AI) will play a growing role in the design and implementation of market over a wide range of applications. The challenges are that how (AI) can guarantee accurate to predict when and why and how consumer behavioral changes to any retail industries. In fact, retailers will need to discover the value that (AI) can bring to what benefits to influence their customer behaviors.

In the future, (AI) will bring their benefits to influence customers to build positive emotions to any retailers in these aspects as below:

1. Future (AI) big data gather tool will be an area of compute science that deals with giving machines , the ability to seem like they have human intelligence. In short, it is the power of a machine to copy intelligent human behavior. For example, machine learning algorithms are being integrated into analytics and customer relationship management platforms to uncover information on how to better serve customers, chat bots have been incorporated into websites to provide immediate service to customers.

2. (AI) adoption continue to rise with chat bots taking the lead. Due to increasing ease of deployment , instant availability and improved quality, chat bots will become more and more common to manage customer service queries and to make intelligent purchase recommendations. Also, retailers can engage this kind of technology to answer continue questions and supplement customer support with chat-based shopping experience. So, (AI) and declines personalized, customized and localized experiences to customers.

(AI) will be applied across the entire retail product and service cycle, firm manufacturing to post-sale customer service interactions. Hence, retailers can use (AI) to its fullest potential will be also to influence purchases in the moment and anticipate future purchases, guiding shoppers towards the right products in a regular and highly personalized manner.

3. (AI) technology can rise the conscious customers. Customers are demanding an increased interest in the ethical practice of the brands they buy from. Todays, customers have a well-developed sense of what is solely intended to drive sales. This has lead to a rise in consumers ho make values based judgements about what to buy and where to shop. These consumers believe their purchase habits have an impact on the world. To win customers, retailers need have good conscious to predict consumers' desire. Future, (AI) data gather technology will be a good consumer behavior predictive tool to predict about for years will now become customer expectations and will have drastically changed the path to purchase. So, (AI) data gather tool is the predictive consumer expectations tool on every interaction, they have these brands.

4. Future (AI) can be impacted to influence consumer behaviors by its potential to free up time, enhance, quality, and enhance personalization. The industries include: Healthcare industry can apply (AI) to support diagnosis by detecting variations in patient data, early identification of potential pandemics, imaging diagnostics; automat industry can apply (AI) to autonomous fleets to ride sharing, semi-autonomous features, such as driver assist, engine monitoring and predictive, autonomous maintenance; financial service industry can apply (AI) to design the suitable personalized financial planning, fraud detection and anti-money laundering and automation of customer operation; transportation and logistics industry can apply (AI) to autonomous trucking and delivery, traffic control and reduced congestion and enhanced security; technology, media and telecommunications industry can apply (AI) to search media, and

recommendation, customized content creation and personalized marketing and advertising to attract retailers to promote; retail and consumer industry can apply (AI) to design personalized production, anticipating customer demand, , inventory and delivery management; energy industry can apply (AI) to read and record smart metering , more efficient grid operation and storage and predictive maintenance; manufacturing industry can apply (AI) to enhance monitoring and auto-correction of processes, supply chain and production optimization and on-demand production.

Hence, future (AI) technology will impact consumer technology when any retailers apply it to assist its manufacturing processes or product sale or service provision processes to satisfy consumers' needs, it means that it can help any retailers to influence positive emotion to consumers in their whole sale or consumption or purchase processes

5. (AI) and machine learning technologies make it possible to capture, process, and inter data on a massive scale effectively , then any human being could ever do. For example, Criteo's creative technology " Kinetic design" can apply insights from 1.2 billion monthly impressions to select and optimize individual branded advertisements components according to each shopper's preference and intent. This ensures more personalization and visually inspiring on brand ads. resulting in up to 12% more sales for (AI) technology advertiser clients.

Moreover, advertisers can now engage and inspire shoppers on a more personal level, rendering custom ads. it real-time for every impression. So, designer continues to learn from each design's success to make ads. more and more effective over time. Furthermore, brands are increasingly using paid search on retail sites to draw attention to their products on the crowded online shelf, e.g. Google shopping is a key growth area's more users are engaging with shopping ads. and across the globe. Google shopping has become essential to retailers' marketing strategies, but is a difficult channel to apply its tool to be promoted effectively . Thus, future (AI) and machine -learning technologies can dramatically improve digital commerce performance application to apply (AI) and machine learning to digital consumer. So, future (AI) technology can be applied to digital commerce aspect, it will fall into the categories of pattern recognition, classification, prediction and consumer behavior.

In conclusion, the benefits of using (AI) in digital commerce include: improved efficiency in discovering the relationships between datasets over traditional methods, which require complex modeling and coding, improved accuracy for clearly defined processes that involve a lot of manual processing, ability to deal with a large emotion of data with many attributes, for example: customer behavior data, multichannel and multi-device data , complex product data and fraud detection, more accurate analysis, such as customer segmentation sentiment, analysis and personalization frequent algorithum refreshes, such as several times a day, to capture the changes in customer and market behavior.

Finally, however, a lot of types predictive consumption behavior around (AI), in particulars that driven by vendors claiming their solutions are (AI) , ready and can deliver dramatic improvements over existing technologies. Application leaders for digital commerce can be misled into believing that (AI) can solve all their problems, which is not true for n in-depth discussion of the (AI) consumers and market behavioral predictive tool and machine -learning technologies bot. Thus, (AI) prediction consumer behavioral technology can give beneficial quantitative analysis for forecasting in business and market especially in consumer behavior and in the consumer decision-making process (consumer choice model) more effectively and efficiently.

Is Artificial Intelligent the most effective and accurate consumer behavioral tool?

Is (AI) the best and the most effective and accurate consumer behavioral prediction tool to compare other kinds of consumer behavioral prediction tools? Nowadays, retailing competitions are serious businessmen often find different kinds of methods to attempt to predict consumer changes. The consumer behavioral predictive methods can include as these below methods, instead of (AI) big data gathering tool.

Firstly, statistics is the popular mathematic method, it applies auto-regression, liner regression, structural equation modelling, logistic regression statistic techniques to be used to predict consumer behaviors. Secondly, it is classification method, it sis a support vector machine to assist businessmen to make consumer behavioral prediction, it also includes decision making tress diagram technique. Thirdly, it is rule mining method, it is algorithm, market base analytic etc. business marketing concept analytical tool, it also includes graph mining technique tool. Next, it is psychological prediction model tool, it is psychology prediction model too, it is a kind of psychological method to predict consumer behaviors. Finally, it is the most updated and potential artificial neural network (ANN) machine tool, it gathered big data, then it will carry on analyzing and applies psychological method to conclude the most accurate and reasonable solutions to give recommendation to businesses to predict when and how and why their consumer behaviors will change. So, it is one owned human mind's machine and owned psychological and analytical efforts to replace humans to make any judgement in order to make the most accurate predictive behavioral changes for consumers, instead of the traditional marketing concept and psychological and mathematic methods to predict consumer behavior, (AI) big data gathering tool will be another new tool.

What are the advantages of (AI) tool to be used to predict consumer behaviors as well as what are the different between it and other traditional consumer behavioral predictive tools? I shall explain as below:

Firstly, as above all case studies are explained to (AI) questionnaire design method benefit, I believe (AI) big data gathering tool can be applied to help human to analyze and design any the suitable valid questions to enquire any kinds of business consumers in order to gather the most meaning and useful opinions to conclude the most accurate consumer behavioral prediction for every questionnaire. So, future (AI)'s analytical effort and decision making effort most be exceed above human's judgement efforts. So, future (AI) can help human to design the most useful and meaning different kinds of valid questionnaire (survey) questions as well as assist humans to analyze and make accurate decision making and conclusions to give opinions to help businessmen to predict when consumer behaviors will change and how their consumption behaviors will change to influence their businesses in order to help them to make any efficient and effective and accurate solutions to avoid consumer number to be decreased and the most important benefit is that it can give opinions to help businessmen to explain why (what the factors) cause their consumer behaviors change suddenly. It will be human's efforts can not achieve to exceed (AI)'s efforts in the future.

Secondly, (AI) can make artificial machine judgement and analytical effort, without human misleading or unfair or unreasonable judgement. So, it can make more fair and reasonable and accurate conclusion to give opinions to predict when, how and why consumer behaviors will change suddenly to the kind of business in customer model building process and evaluating the results of customer relationship management –related investment more accurate.

Furthermore, (AI) big data gathering tool will help businesses to improve the success rate of acquiring customers, increasing sales and establishing competitiveness. (AI) big data gathering tool can give opinions how to build customer loyalty to be positive emotion impact and it can find solutions to avoid every client's negative emotion causes to bring complaints behavior to the businessman's product or service. For example, Telecom industry and aggressive research has been conducted in this by applying various data mining techniques to avoid long distance phone call users' complaints. If gathered any long distance phone call users' past complaint data to record what are their general complaint issues. Then, (AI) tool will analyze all these past complaint issues to conclude and give opinions to let Telecom knows whether which aspects encounter challenge that Telecom needs to improve it's long distance phone call services or functions in order to satisfy Telecom's long distance phone call users' needs for long term. After Telecom attempted to improve its services and/ or functions from (AI) opinions and solution methods, when it fell it's long distance phone call users have positive emotions to satisfy its service performance and function performance. Then, it can prove (AI) tool's opinions and solutions are useful. The consequence is that their complain numbers will be decreased and they won't plan to choose another long distance phone call telephone service company to replace Telecom long distance phone call service more easily.

So, (AI) big data gathering tool can concentrate on finding focus on components of customer relationship management method and datasets more accurate and efficient and effective than human's data gathering and analytical effort. It implies (AI) big data gathering tool has unique more efficient and effective and accurate dataset gathering and analytical and judgement and decision making effort, it is human can not achieve.

Thirdly, (AI) big data gathering tool has much customer loyalty predictive effort. It's effort is more easily subsequently selected, reviewed and classified to compare human's gathering data effort in whole data gathering and analytical process.

In (AI) big data gathering process, (AI) can organize whole big data gathering process and technique more easily in short time. It will include these four steps. The first stage is that customer identification stage, customer identification also known as acquisition has to do with targeting the population , who are most likely to become customer segmentation. So, (AI) can help different kinds of businesses to gather their competitors' consumer purchase behavior data in short time, it is human can not achieve.

The second stage is that customer attraction stage, after (AI) maker has been segmented for the business when it has ensured to gather the businessman's global competitors' consumers data. Then it analyze these all data to find solutions / methods to give the best opinions to the organizations how to achieve the direct effort and resources into attracting the target customer segments. The third stage is that customer retention, it can be defined as the activity that an organization undertakes in order to reduce customer defections. TO be successful, customer retention starts with the first contact on organization has with a customer and continues throughout the entire lifetime of a relationship involves loyalty programs, one to one marketing and complaints management. SO, (AI) can consist the business to find the best or the most reasonable , efficient , effective solutions or methods and it will conclude all these solutions to find the most reasonable and useful opinions to achieve to the aim to help the business to reduce customer complain numbers and help the business to build confident loyalty relationship between it and its clients. SO, (AI)'s analytical effort and decision making effort can be more accurate than human's analytical effort and decision making effort. IT can achieve it's consumer behavioral predictive aim more accurate and efficient and effective in the shortest time to compare human.

Fourthly, (AI) big data gathering tool can design more accurate dataset program for questionnaire (survey) to compare human's questionnaire (survey) effort. It means that (AI) can spend less time to research and make judgement what are the most reasonable and meaning questions for different kinds of businesses' needs. This includes data conduction a questionnaire, survey or interview of the individual or environment researched, public data repository: This includes commercially available public data; organizational data; this contains data collected from an organizational database, organizational information system. For example, their website log details etc. It also includes company transactional data, data purchased from a company.

For example, one vehicle sale company expects to research all global vehicle sale companies' past the different kinds of vehicle styles, design sale number data, the different kinds of vehicle style, design sale price data, every country's vehicle consumer number to the vehicle purchase number data to the vehicle company in short time. (AI) big data gathering tool can help the vehicle sale company to gather all any one for these global vehicle sale competitors' past data in the short time. It is human effort, who can not achieve this efficient, effective and accurate data gathering aim for this vehicle sale company. Even, when (AI) had gathered all global it's vehicle competitors' past sale data, (AI) can make more accurate analytical and judgement and decision making effort to design different kinds of questionnaire (survey) questions to prepare to enquire it's different target segmentation vehicle potential clients in order to predict what are their needs to choose to buy any vehicles from the vehicle company. SO, (AI) tool can conclude more accurate conclusions and give the most reasonable and useful opinions to let the vehicle company to know in order to predict what are it's potential vehicle buyer's needs and manufacture the suitable vehicle styles or designs to raise their vehicle purchase desires.

Fifthly, (AI) tool is only one perfect tool for big data gathering in order to achieve accurate results and increased profit. What is (AI) big data gathering mean? The term " big data" gathering describes the accumulation and analytical of vast amounts of information, but big data is much more than a big amount of data. It is also the ability to extract meaning to sort through big volumes of numbers and find the hidden patterns, unexpected correlations and surprising connections that can be used in different industries like medical field, security and protection field or marketing that adopt " big data driven" decision making enjoy significantly greater productivity than those that do not. So, the benefits of (AI) is given to the company by using big data repaid complexity of implementation projects and hence project risks, when accelerating time to value. It is why that human's gathering effort can not replace (A I) data gathering effort.

All analysing above benefits to (AI) big data benefits to any organizations, it brief this question: How can (AI)apply big data gathering and analysing to predict when and how any why consumer behavior will change suddenly? The purchase decision making process is consumers reducing purchase choice behaviors.

Consumers are being considered pure rational beings (consumer tried only to satisfy self-interest). Hence, due to future (AI) owns human's psychological , analytical , emotional predictive, purchasing decision making effort.

(A I) will be assumed to sees one customer how who will make purchase decisions. So, after the (AI) gathered all data concerns the find of business's past customer segmentation purchase activities, e.g. age, sex,. Income level, the product's style sale number, the product price variable sale etc. different kinds complex data.

It can make more accurate psychological and analytical effort to predict when the business's consumer behaviors will change as behaviors will change as well as find what reasons their consumption behaviors will change and how trend of their consumer behaviors will change more accurate. For example, today there are a lot of industries that use big data: healthcare (treatment) becoming personalized and patient centric and predictive analysis are used to prevent diseases for example Angelina Jolie under event a predictive double mastectomy after learning she had 87% rich to developing breast cancer, sports (by using sensors data are collected from players during a game in order to improve their playing schemes), weather(more than 60 years of global weather analysis are used to predict the risk of future extreme events), logistics (smart tucks and smart species, agriculture (monitoring weather and soil conditions for optimum point of harvesting).

Consequently, due to the evolving consumer demands, and the ever growing digitization, the world is digitally transforming which means the new technologies are needed to be used and driven significant business improvement. So, such as why (AI) tool will be our future main predictive tool to help businesses to predict when, how and why their potential customer behavioral will change. Big data is one of the our channels through digital transformation is made, together with cloud, mobile and networks. The challenges for digital transforming and therefore using A I big data gathering tool as main technology are: digital proficiency, legacy systems, security and jobs becoming absolute.

In the future, big data can use data from text to picture , sounds, movies, music satellite coordinates or any other type of input or output data that type of input or output data that came from different influential aspect. It is cloud solutions, bring big data will be for predict insight driven by business strategy, new product strategies and new consumer relationship, predictive consumer behavioral strategy. Using the right data in the right business decision will mean smart decisions, new opportunities and utimately a big competitive advantage. Hence (AI) big data gathering tool is different is that (AI) can be one depth in-memory database function, it can make real-time data analytics that provide meaningful information in short time, it is also the visualization tool , such as SAP Lumira, allow this exploration and understanding of the data, and ultimately supports the decision making process. All above these features, which will be human's data gathering effort who won't exceed (AI) big data gathering effort. Hence, future (AI)big data gathering will be the best choice to assist businesses to predict consumer behaviors successfully.

Reference

Adrian, P. (2012). Introduction to marketing theory & practice, 3 rd edition, London: Oxford press.

Ajzen, I (1991). The theory of planned behavior. Organizational behavior and human decision processes, 50(2), 179-211. doi: 10.1016/0749.5978 (91) 90020-7.

Alba, Joseph W. and J. Wesley Hutchinson (1987). " Dimensions Of Consumer Expertise", Journal of consumer research, 13 March, 411-454.

Bailey, L., Mokhtarian, P.L. Little, A. (2008). The broader Connection Between Public Transportation, Energy Conservation And Greenhouse Gas Reduction, Report Prepared As Part Of TCRP Project J-11/Tasks Transit Cooperative Research Program, Transportation Research Board Submitted To American Public Transportation Association in http://www.apta.com/research/into/online/land_use.cfmi, accessed 17 April 2008.

Baucer, R,"Consumer Bhavior As Risk Taking , In Risk Taking And Information handling In Consumer Behavior", D. Coxceds Harvard University Press, Cambridge, Mass 1976.

Biederman, P. (2008). Travel and tourism, Pearson Prentice Hall, New Jersey.

Bogers, R. P., Brug, J. Van Assema, P., & Dagnetie, P.C. (2004) , Explaining fruit and vegetable consumption: The theory of planned behavior and misconception of personal intake level. Appetite, 42,157-166.

Bolton, Ruth N. (1998), " A Dynamic Model Of The Duration Of The Customer's Relationship With A Continuous Service Provider: The Role Of Satisfaction", Marketing Science, 17 (1), 45-65.

B.Shiv and A. Fedorikhin, " Heart And Min In Conflict: The Interplay Of affect And Cognition In Consumer Decision Making", J. Consumer Res., vol. 26, pp. 278-292, Dec. 1999.

Brown, K.W., Ryan, R.M. Reswell , J.D. (2007). Mindfulness: Theoretical Foundatins And Evidence For Its Salutary Effects. Psychological Inquiry, 18, 211-237.

Burke, R.R. : Behavioral effects of digital signage, J. Advertising Res. 49(2), 180-185 (2009).

Cant, M., Brink , A. & Brijall, S., Consumer behavior, Cape Town, South Africa: Juta, 2006.

Conner, M. & Abraham, C. (2001). Conscientiousness and the theory of planned behavior: Toward a more complete model of the antecedents of intention and behavior. Social psychology bulletin, 27, 1547-1561.

Cooper C. Mallon, K, Leadbetter S, Pollack L, Peipins (2005) , cancer internet search activity on a major search engine, United States 2001 to 2003, J Med Internet Res. 7(3): e36.

Cope, R. R. Cope and H. Davis (2008). Disney's virtual Queues: A strategic opportunity to co-brand services ? Journal of Business & economics research, vol. 6 no10, 13-20.

Cornelia, B.F. (1999) Rural development news, the North Central Regional Center For Rural Development vol. no 24 , IOWA.

Couper, M.P. J. Blair and T. Triplet (1999). A Comparison Of Mail And E-mail For a Survey Of Employees In USA Statistical Agencies. Journal Of Official Statistics, 15, 39-56.

David J. Nowak & Gordon M. Melsler (2016) " Air quality effects of urban trees and parks." National recreation and park association, USA.

Data monitor (2008). The proctor and gamble company. Retrieved Nov. 15 2009 from http://www.datamonitor.com/

De Hollander, A. E. M., J.M. Melse, Elebret & P. G.N. Kramers (1999), " An Aggregate public health indicator to represent the impact of multiple environmental exposures" Epidemiology: 606-617.

De Visser, R.O., & McDonnell, E.J. (2013). " Man points": Masculine capital and young men's health. Health psychology, 32(1), 5-14. doi:10. 1037/ a0029045.

Dunn, J & A Neumsister (2002). Knowledge management in the Information age. E. business review, Fall , 37-45. Jounral of service, spring 2011, vol. 4, no1, De Grovte (2009).

Dyer, D., F. Dalzell & R. Olegario (2004). Rising tide. Lessons learned from 165 years of brand building at Procter and Gamble. Boston, MA: Havard Business School Press.

Eysenbach G (2006) Infodemiology: Tracking flu- related searches on the web for syndromic surveillance. American Medical Informatics Associaion Annual Symposium Proceedings , Curran Associates, Red Hook, NY, pp. 244-248.

Ettredge M, Gerdes, J. Karuga , G (2005) Using web- based search data to predict macro-economic statistics. Commun ACM 48: 87-92.

Felce, D. and Perry, J. (1995). Quality of life: A contribution to its definition and measurement, vol. 16, no.1 pp: 51-74.

Feldman, Jack M. And John G. Lynch Jr. (1988), "Self-
Generated Validity And Other Effects Of Measurement On Belife, Attitude, Intention And Behavior", Journal of applied psychology, 73(3),421-35.

Fiese, M, Hofmann, W., & Wanke, M (2009). The impulsive consumer. Predicting consumer behavior with implicit reaction time measurement. In M. Wanke (ed.) Social psychology of consumer behavior (pp.335-364). New York, NY: Psychology press.

Fitzsimons, Gavan, J. And Vicki G. Morwitz (1996), " The Effect Of Measuring Intent On Brand-Level
Purchase Behavior", Journal of consumer research, 23 (1), 1-11.

Hallerman , D. (2008) video Advertising Online: Spending And Pricing , New York. E-Marketer.

Harriet Griffey. (2010) The art of concentration, enhance focus, Reduce, stress and achieve move. Macmillan publishers ltd,Basinastoke and Oxford, London UK.

Helleman, D. (2008) Video Advertising Online: Spending And Pricing , New York, E-Marketer.

Hensen, C. (2003). Kreuzfahrtourismus.www.christoph- hensen.de/ Facharbeit.pdf.

Huang, H.I. (2012). An empirical analysis of the strategic Management of competitive advantage: a case study of higher technical and vocational education in Taiwan (Doctoral dissertation,
Victoria University).

Jamieson, Linda F. And Frank M. Bass (1989), " Adjusting Stated Intention Measures To Predict Trial Purchase Of New Products: A Comparison Of Models And Methods," Journal of marketing research, 26 (August), 336-45.

Korea Ministry Of Environment. Public Organizations spend 2.2 Trillon

Korean Won To Purchase green Products in 2014; Ministry Of Environment: Sejoung, Korea, 2015.

Kremers, S.P. J., De Bruijn, G.J., droomers, M., Van Lenthe, F. J., & Brug, J. (2005). Environmental interventions for selected dietary behaviors in adults. In J. Brug & F. J. Van Lenthe (eds.) , Environmental determinants and interventions for physical activity, nutrition and smoking: A review pp. 282-315. Rotterdam: Erasmus Medical Center.

Lee, D.; Kim, M. ; Lee, J. adoption of green electricity policies: Investigating the role of environmental attitudes via big data-driven search-queries. Energy policy 2016. 90, 187-201.

Lee, Terrence, " Tech in Asia-connecting Asia's startup system " Tech. in Asia-connecting Asia's startup ecosystem, N.p.,4 July 2016.

Los Angeles Country Department Of public Health (2016), Country Health Ranking Model, Retrieved From www.countryhealthrankgings.org/our-approach. USA.

Mayne, Lonnie. " Evolve of die in the age of the consumer". Entrepreneur, N.P. , 16 Apr. 2014. web of Oct. 2016.

McGregor, S.L. T., & Goldsmith, E.B. (1998). Expanding our understanding of quality of life, standard of living and well-being. Journal of family and consumer science, 90(2), 2-6, 22.

McMichael, A.J. M. Mckee, J. Shkolnikov and T. Valkanen (2004), " Morality trends and setbacks, global convergence or divergence?", Lancet 363, 1155-1159.

Melse, J.M. & A.E. M. De Hollander (2001). " Human Health And The Environment", background document for the OECD Environmental Outlook, OECD, Paris.

Moschis, George p. & Roy, L. Moore (1979), " Decision making among the young. A socialization perspective " Journal of consumer research , 6 (September).

Mulligan, M. Banerjee, T & Thomas, N. (2008) ,European Paid Content And Activity Forecast, (2008 to 2013), Jupiter Research.

Peter, J., Ryan, M, M, " An Investigation Of Perceived Risk At The Brand Level, " Journal of marketing research, 13 May 1976, pp. 184-188.

Pieters, R., & Wedel, M. (2007). Goal Control Of Visual Attention To Advertising: The Yarbus Implication. Journal Of Consumer Research, 34, 224-233 (August).

Parasuaman, and Leonard L. Berry (1985), " Problems And Strategies In Sevices Marketing", Journal of marketing, 49 (Spring), 33-46.

Priesnitz, W. (2007) Counting Our Food Miles. Natural Life, 1 July.

R.C. Oliver, " When is consumer loyalty?" J.Marketing vol. 63, pp.33-44.1999.

Reggiani, A . (ed). 1998, accessibility, trade and locational behavior, Ashgate publishing ltd, England.

Rushe, D. (2013) " The 10 best paid CEO in America". The Guardian , 22 Oct, (online). Available at: http://www.theguardian.com/business/2013/Oct22/best-paid-chief-executives-america (Accessed: 3 May 2014).

Spiekermann and Wegener (2007), update of selected potential accessibility indicators. Final report, urban and regional research (S&W), RRG spatial planning and geoinformation. ESPON. Available online at http:// <www.espon.eu/mmp/online/website/ contentprojects/947/1297/ file_2724/espon_accessibility_update-2006-fr_070207.pdf>, accessed on 1 July 2009.

Starbucks (2014) Our company available at http:// www. starbucks.com/ about- us/company-information (accessed: 3 May 2014).

Shostack, G. Lynn (1984), " Designing Services That Deliver", Harvard Business Review, 62 (January-February), 133-9.

Shostack, G. Lynn (1985), " Planning The Service Encounter ,in the service encounter" , John A. Czepiel, Michael R. Solomon, and Carol F. Suprenant, eds. New York: Lexington Books, 243-54.

Shostack, G. Lynn (1987), " Service Positioning Through, Structural Change", Journal of marketing, 51 (Janurary), 34-43.

Soloman, Michael R. (1985), "Packaging The Service Provider", Service Industries Journal , 5(1), 64-71.

Stevens, C.W. (1980), "K-MartStores Try New Look To Invite More Spending" The Wall Street Journal, Nov. 26, 29-35.

Sullivan, Nicholas P(2007). You can hear me now: How Micro loans and cell phones are connecting the world, San Francisco, CA: John Wilsey & Sans, 2007.

T. Ambler, A. Ioannides, And S. Rose, " Brand s On The Brain : Neuroimages Of Advertising ", Business Strategy rev., vol. 11, 3. pp. 17-30. 2000.

Westbrook, Robert A. (1980), " Intrapersonal affective influences on consumer satisfaction with products, " Journal of consumer research , 7 (June) 49-54.

Wiig, k.(1993). Knowledge management foundations: Thinking About thinking. How people and organizations create, represent and use knowledge vol.1 , of knowledge management series schema press: Arlington, TX.

World Health Organization (2003). Diet, nutrition and the prevention of Chronic diseases report of a joint WHO/FAO. expert consultation. Geneva: World Health Organization.

Wysocki, B. (1979), " Sight, Smell, Sound: They're all arms in retailer's arsenal" The Wall Street Journal, Nov. 17, 1979. 1-35.

Yale Center For Environmental Law And Policy (2006). Environmental Performance Index. Data available on-line at http://epi.yale.edu

FOUR

WHICH KINDS OF INDUSTRIES WILL BE INFLUENCED BY FUTURE ROBOTS TECHNOLOGICAL DEVELOPMENT TO BRING ECONOMY GROWTH

What (AI) technological development will influence what kinds of UK and US industries development within ten years? Are environment and education and automatic manufacturing technologies will be UK and US future (AI) new technological development trends? What will be the difference between the (AI) developed countries and (non AI) developing countries future technologies development both in the future?

1 (AI) online teaching technology development

Future, (AI) online teaching method will be popular to be applied to teach to any university in possible, even secondary and primary schools. Because internet service is free charge to any students in any countries. Many different age students who can know how to apply internet as well as internet studying is very convenient to any students who can to internet to learn or study in home or public library or school library conveniently. Teachers do not need spend much time to teach students in classroom. They can use internet to teach teachers by face to face seeing and talking to their individual student from every student's computer. So, students do not also often spend much time to go to school to learn. So, developing any fast speed and time saving and talking and listening online teaching methods will be popular needs to any UK primary and secondary and university students in the future. It will be one new technological teaching method to change the traditional classroom educational method in UK and schools. For example, when one UK student who had left UK and is living in another country long time. If any UK school did not provide online teaching service to any UK students. It means that the UK citizen can not choose study himself/herself any UK school if who still hope to study any UK course when who is living in another country. Even one foreign student who does not go to UK to study, if he/she can find any UK primary or secondary or university to study from online. Then, the UK school won't lose one foreign student, due to it does not provide online teaching method to any foreign students. So, online Technology educational learning method will be one popular learning method which is enhanced, supported, mediated or assessed by the use of electronic media. Technology also enhanced learning may involve the use of new or established technology and/or the creation of new learning material. It may be deployed both locally and at a distance (i.e. a combination of traditional and e-learning approaches), to learning that is delivered entirely online. Online learning technology characteristics (features) include identification of a project lead for each area of any learning strategy, identification of two " quick win" for example lecture capture, electronic submission and feedback.

How can online technology enhance learning at UK any schools? It will include these several aspects to analyze. On identifying, prioritizing and innovation hand, online technology is a process for resourcing, prioritizing, acquiring and evaluating school software and hardware for UK any school needs. On staff development learning plan and a student skills development plan hand, UK schools need to establish a base-line policy on the standard (minimum) technology enhanced learning expectation for education each program and module and a mechanism for updating the schools‘ policies. On evaluation and research hand, a mechanism for engaging the owners of the technology enhanced learning strategy with best practice in the sector including contributing to and benefiting from pedagogical research and the evaluation of the student experience to UK any school.

Thus, UK schools can apply (AI) teachers to teach their students from online technological teaching channel to develop on educational aspect, such as (AI) teachers' digital literacies and appropriate technical skills that equip UK students for life-long learning, graduate level employment and professional practice, be empowered to learn how to learn with online teaching technology, using online technology to engage in interactive, creative and co-constructed learning with the potential for online learning in an interdisciplinary and international context, using online teaching technology to engage in learning with and from people from anywhere in the world, be supported on placement and in workplace learning through mobile applications and other supportive technologies that facilitate their online learning when away from the classroom, having access to innovative methods of online learning teaching and assessment that are the foundation of a research-lead academic environment, engaging with UK schools in developing , implementing and reviewing the technology enhanced learning strategy. Thus, in the future, it is important to build a capacity to apply (AI) teaching robots to teach their students from the online education technology to adopt future learning innovation and student individual online learning need (demand) to UK any school (AI) robotic online teaching trend.

There are many examples where UK academics working in isolation or in small UK teaching organizations or classroom learning groups have developed (AI) robotic teaching innovation that have a positive impact on UK students‘ academic experience , but these have remained isolated to particular modules or occasionally program. The aim of education researching online learning process is to identify the good (AI) robotic teachers' online teaching innovation that is being developed and to prioritize those that have the potential to make a significant contribution to improving the academic student experience at UK any schools. This online teaching process will need any UK schools which can plan how to apply limited resources necessary to achieve online teaching. In addition, the online teaching research process would evaluate and prioritize large scale educational software and hardware requests for primary, secondary and university students‘ requests. An important part to this process will be to ensure the integration of (AI) robotic teaching tools and their educational method to be applied to online educational products and packages that school staff and students regular use to make routine working and access as seamless as possible.

Decisions about school administrative online technologies should not be taken in isolation before assessing the impact on UK teaching staff. In addition, a range of techniques such as, (AI) robotic online expert facilitation, (AI) robotic coaching and peer support will be used to support individuals, groups or longer academic units, who are learning on major technology enhanced (AI) robotic teching online learning projects. Staff engagement may also facilitated through incorporating technology that is used in teaching staff research and/or professional activity that can be cooperated into their teaching.

Consequently, (AI) robotic online learning technology can develop UK students skills, UK schools need to understand how UK students understand technology and learn with it, therefore the digital literacy strategy needs to be considered as part of the overall strategy as well as the relevant skills development in UK employability strategy. So, in the future, (AI) teaching robotic online learning and teaching technology will make it clear that students will develop technical skills the appropriate level for graduate employability and professional practice. Also, in the future, the (AI) robotic teaching online technology can enhance learning working group to discuss external development, that are of educational strategic importance, understanding and evaluating current best practice and research and understanding and evaluating the online educational strategic contribution that pedagogical research and student feedback can have on online educational strategy, policy and practice. The (AI) robotic teaching e-learning unit is responsible for informing and educating. This could be done by, for example, providing a short digest of relevant information for each meeting and by setting aside a proportion of each school meeting to discuss a topic of particular (AI) robotic teachers to be applied to online educational strategic interest to every school. Academics that have not got a specialist interest in (AI) robotic teaching online educational technology enhanced learning will need relevant information at an appropriate time. This could be provided at a school department or faculty level and this will have clear links to the staff and (AI) robotic teaching online teaching development plan. Hence, future (AI) robotic teaching online educational development strategy will influence any UK or US educational school technological improvement in the future (AI) robotic online teaching method.

2 (AI) robotic environmental protection technology

Can future (AI) robotic environment technology be valid to human to develop? Nowadays, global air and water pollution is serious. For example, UK has many farming is polluted by the water and air pollution. It will influence UK farmers' income if whose farm land (natural resource) is polluted by water or air (natural resource). Even it will influence UK citizen will encounter food shortage if UK farmers can not grow any fresh and health food to provide the enough food numbers to eat every day. Moreover, air and water pollution will influence UK citizen drink the polluted water and breathe the dirty air to live every day. This natural resource (air and water challenge) will influence UK citizen health to cause illness , even death every easily. So, UK government can not neglect the natural environment pollution challenge. The environmental protection technology will help the UK and development countries to solve the challenge of climate change to avoid or reduce farming, foods, or vegetable or fruits or rice, pork, livestock numbers loss threats, i.e. the development and deployment of low carbon energy technology, including technology for the efficient use of energy. The commercialization of low carbon energy and energy efficiency technologies in the UK, with a specific focus on the demonstration and deployment phases of bringing low carbon technologies to UK market.

The UK Government needs to deliver a low carbon economy and to meet UK ambitions emission reduction target. So, low carbon and environmental protection technology researching and development will reduce the carbon intensity of energy production as well as reduce energy demand, towards meeting the contributing UK's ambitions production as well as reduce energy demand, and renewable energy goals. The use of energy (including transportation fuel) and the UK's targets on climate change, for example, by helping the UK make a step change in increasing deployment of renewable energy, improving UK energy efficiency and helping low carbon technologies reach the market. The development of low carbon technologies, and to realize the benefits of doing so in terms ensuring security of energy supply for the UK future economy development.

In UK, private sector investment in technology innovation in the low carbon energy sector will other sectors of the economy. So, in UK energy technologies are likely needed to be developed to avoid dangerous climate change, or an acceptable cost. So, in the future, UK government will need to consider to research environment protection and low carbon energy technology. The activities will reduce carbon emissions, or have the potential to reduce carbon emissions on the longer term, through the use of energy technology will accelerate development and deployment of low-carbon energy and energy efficiency technologies will capacity in the demonstration and deployment of low carbon technologies. Innovation in the energy sector is the only way to identify, develop and reduce the costs of new and improved technologies for the extractions, generation, distribution and use of energy. It has long been an important means of achieving the UK's energy policy aims of a secure and affordable energy supply, as well as to develop the environmentally friendly technologies that are required in UK response to climate change, i.e. nuclear, wind or water, sun energy technology, which is future new energy technology is suitable to research to create to apply instead of current electricity energy.

How global warming influences UK agriculture growth. Scientists have also been fighting the use of chlorine in municipal water systems to kill various strands of bacteria. Chlorine reduces by about 80% the number of alimentary tract diseases relative to polluted, unchlorinated water. A relatively new genetically modified agricultural products. They were partly successful in Europe, such as UK (some countries banned genetically modified products) in spite of the fact that neither history nor research supports their case. People began to modify plants as early as the beginning of the agricultural revolution (8000 to 10,000 years ago), when they started seed selection and who have continued ever since. The green revolution of the 1960 year brought about strains of grans and rice more resistant to a variety of local conditions. The effects have been that countries like India, which had suffered from recurrent famines over the millennia, became self-sufficient in food due to the resultant sharp increase in agricultural productivity. It was a real science and technology over the poverty dominating most of human history. But it is precisely the products of science and technology that ecologists are so deathly afraid of. In an interesting study in a quarter (28%) of clinically analyzed cases of obsessive compulsive disorder were cases resulting from the fear of global warming.

To destroy the modern, whether industrial or postindustrial, civilization, human have to destroy an important engine of economic growth, that is its energy sources. And this is what eco-warriors try to achieve under the banner of against global warming. Thus, UK government will have responsibility to attempt to research new technology to fight global warming challenge for itself farmer benefits and even global benefits both on the future.

Hence, future (AI) robotic development can be applied to environment protection aspect. Future (AI) robotic tools can help human to predict when and how any why environment pollution will occur in which countries and (AI) robotic tools can be one environment protection machine to gather environment pollution information to give opinions to human how we ought need to do in anywhere in order to avoid the places' environment pollution will become serious in influence our health. So, future (AI) robotic machince will be one predictive environmental pollution and bad climate change machine and it can give opinions to avoid serious environment pollution and give solutions to solve environment pollution any country.

(AI) will give global warming technological protection economic influence opinion to human

Some future economists indicated reasons to explain why UK government and businessmen needed to consider how to develop natural environment protection technology to avoid global warming challenge to influence UK economy development. They indicated the anthropogenic (human-made) global warming resulting from the increase in "greenhouse gas". They offered their perspectives on the scientific valid of anthropogenic global warming phenomenon, its probability of occur and expected consequences and is dominated by technologists, economists and political scientists, who considered the need to make the horribly costly adjustments in energy generation and usage suggested by climate alarmists.

Many stress that global warming is primarily caused by other phenomena than human use of fossil fuels or human activities in general. They are looking at the activities of the sun and impact of the larger universe as the main source of global warming and stress that global warmings (plural) happen intermittently with global cooling. I shall explain why global climate warming will influence to the political and economics of the issue to UK country. For it is the latter, rather than the global warming itself, that will pose a challenge to the Western world, such as UK and the world at large in the future. Scientists concerned who should move forward with policy measures to avert the alleged disaster. They also apply manufacturing theories to support enough to frighten politicians into action and scare societies into acceptance of measures that would sharply reduce UK citizen their living standards. Otherwise, UK politicians had support that bureaucracies were established, money allocated and lobbies created dependent on the new kind of subsidies. In consequences, climate alarmism and resultant interventions in national economies and human activities have become the increasingly wide spread and increasingly cost reality. With the growing availability of money distributed, and even more promised, a range of benefit of the global warming machinery has been on the increase. So, if UK government did not concern how to innovate new weather protection technology to avoid climate change adverse (poor) influence. It is possible that billions of dollars of UK public money are needed to spend on research global warming challenge because global warming will influence UK agricultural industry. UK agricultural industry is one important export income source to raise UK GDP income every year. If global warming become very serious to influence UK weather to be bad to cause UK farmers who can not grow good taste food and vegetable to supply to domestic and overseas food consumers to eat. Then, UK will loss much GDP income from local agricultural export sale. It seems global warming and agricultural production which has direct relationship to influence UK economy development in the future.

The main problem with climatology is that it must be based as already stressed on very many variables affecting climate and too few hard data necessity. Differences apply not only with respect to the scale of changes obtained, but even to their direction (rising or declining temperature). Some weather scientists indicated to concern global warming challenge. In consequence, it would be impossible to discover if and where errors were made not only in estimating relationships between variables but also in the quality of data used. (Hauser, J. Tellis, G. J; Griffin, A. 2006) They were comparing average temperatures measured some 30, 40 or 50 years ago by, say, 90 % weather stations in the countryside and 10 % stations in the cities with contemporary average temperatures measured by weather stations located today on 50:50 basis in the countryside and cities. Then, one could obtain the increasing temperatures without any real world climate or even weather changes. Comparability would be ensured if the same number of countryside-located and city-located weather stations had been compared for different periods. The alarmists intentionally mix up " temperature growth" with the trend of temperature growth. To give an example, if in the first decade the temperature grew by 0.5 % degree, in second decade it grew by 0.3 % and in third decade it grew by 0.1%, what was registered was a growth in the temperature, but certainly not a trend of growing temperature. A fourth decade should, on the basis of the trend, bring about

no change in the temperature.
To conclude, scientists believed that global warming was caused by human's bad behavior more than natural environment influence. So, it is human's responsibility needs to solve this challenge, due to who feel earning profit aim is more important to protect natural environment, e.g. air and water pollution , due to manufacturing process is the main factor. So, UK has responsibility to attempt to research how to solve global warming challenge , such as it has many famous scientists who can devote their scientific skills to cooperate to solve global warming challenge with other countries' scientists. Some weather scientists also hypothesized that human may be at the end of the present warming period. If they are right, it would be bad for humanity, as warmer periods have always been associated with better conditions for economic activity. To sum up, scientists believed that global warming will influence human economic activity to be bad.

Future (AI) robotic environmental protection machine can give opinions to UK farmers:
Climate alarmists were able to convince a large part of the Western public and a majority of Western politicians of the cause of fighting against the global warming. It supposes itself in an instinctive preference for collectivist solutions in economic and social spheres, with negative to disastrous consequences when scientists are applied in practice, so UK government needs to concern global warming challenge, due to it is possible that it will influence UK natural environment weather to be poor to influence many UK farmers' agricultural and vegetable and fruit and rice wheat etc. food growth successfully. What is the global warming influence to cause disease? For example, ecological alarmists and activists (eco-warriors) never admit they are wrong, they long pursued their fear mongering campaign against chlorine. Their success in branding DDT a dangerous substance had a negative impact on the malaria eradication campaign in poorer parts of the world. Alternatives to be have been far less effective and the result has been the resurgence of malaria cases and the manifold increase in malaria -caused deaths to the largest extent in Africa.

3 (AI) robotic automation technology in manufacturing industry
Future, (AI) robotic automation technology can be applied to manufacturing industry. For example, nowadays, UK computer and space explore technology had reached the mature stage. It means that UK government ought not need to continue spend much resource to research these two kind technologies. Otherwise, the (AI)robotic automatic manufacturing

technology, e.g. human intelligence new product. It has need to develop because human intelligence machines will bring beneficial to satisfy human everyday life need, e.g. hospital patients' activities need, if the patent who can not walk easily, but the human intelligence machine can assist the patient walk to anywhere conveniently. So, he/she does not need to sit on wheel chair and apply the human intelligence machine man to help him/her to drive on the intelligence automatic driving vehicle to go to anywhere conveniently.

Otherwise, increased automation in low wage countries, e.g. China, Korea, Africa, Hong Kong etc. which have traditionally manufacturing firms, could use automatic technological manufacturing to bring lose cost advantage and potentially lose their ability of achieving rapid economy growth by shifting workers to factory jobs. So, UK government and businessmen needs to consider automation technology development, i.e. 3D printing manufacturing industry will encourage UK companies to move manufacturing process, closer to gain the biggest advantage from this 3D automation technology development.

A growing concern of premature de-industrialization in energy and developing countries could require new models and a need un-skillful the UK workforce. In the future, the best way toward for UK cities will reduce their exposure to automation is to boost their technological dynamic and attract more UK skilled workers. Automation technology progress can give UK manufacturers' employee benefits, such as long term healthy productivity improvement, raising productivity efficiency and product quality, macroeconomic and microeconomic effects of automation technological change, it's change will be beneficial to UK society, i.e. automation active labor market policies, which could help UK job seekers find jobs from training to incentive to support self-employment to create high technological job employment chance in UK society. So, raising science, technology, engineering and math subjects update skills level are needed to UK any universities, which can be increasingly important in UK society, these factors could complicate the ability of UK high automation technology education to adopt to the UK automation manufacturing technological change. A talent mismatch already exists in UK, with many well UK educated workers can find employment in lower-skilled jobs. To combat this, greater coordination will be needed between the education, training and employment sectors in UK society.

Why are high automatic technology product development models needed to research to UK any manufacturers? UK government and manufacturers need to consider how to achieve high technology product development models. According to Hauser et al. (2006) indicated the high technology (high tech.) development process, is influenced by the innovative process, bringing products on exception value which stimulate product market demand. Innovation provides products the specific basis for which world economies compete with each other on the global market. Able to find new solutions, innovations generate significant changes in existing markets, destroy them, or create new marketing (Hauser et al. 2006). So, UK manufacturers need to concern on any manufacturing high technology product development process because which can influence any new products development to manufacture to sell to any overseas or domestic both markets successfully.

What is high tech. product meaning? Mohr et al. (2010) argues that there are two reasons why it is important to clarify and specific high technology : (1) due to the impact of technologies on the economy, attempts are made to classify economic production and incomes ; (2) due to the impact of high tech. on the environment. Standard marketing strategies are being modified and adopted , therefore, it is necessary to know the products to focus on. Why UK manufacturers need to consider high technological product process. Nowadays, high tech. products are complex, advanced, requiring specific technical knowledge, which is technologically not discontinued and being produced at the companies which have twice as many technical personnel and invest twice as many in scientific research and development than other companies. Moreover, these products are time-sensitive as scientists are continuously searching for new approaches for invention of more advanced technologies which make all preceding ones lower-ranking. The most important, nowadays global consumers will adopt the particular technology. It means that global customers may delay adopting new high-tech. products and in order to mitigate the prolonged uncertainty require a high degree of education and information about the product and need post-purchase reassurance.

Anyway, nowadays customer individual needs in high tech. environments are characterized by sudden changes related to unpredictable fashion. Even, consumers concern about how to preserve new product‘ competitive technological standard is completely incompatible with technological uncertainty. The most important factor is the prevalence rate of any new products development process, which is influenced by slower than of traditional products. In many cases high-tech. automatic product market are being materialized slower than which are expected. The technological uncertainty challenges will exist in development process, such as uncertainty related to the timetable for development of the question whether the new product will be function as promised. In automatic high-tech. industries, the time requires for product development is difficult to predict as , commonly, it takes longer than expected , uncertainty related to unanticipated consequences and uncertainty about the product life cycle related to competition products. In conclusion, these factors will influence new automatic technology product development process unsuccessful, so UK manufacturers will need to concern on any high technological automatic product's manufacturing process.

Future economists predict automatic technology how to influence future UK economy

Before, all over the world presented picture of demonstrate in London on the occasion of the meeting of the G20. Some economists indicated disastrous economy consequences will occur to any one of Western country , such as UK, so if any one of Western country did not consider automatic technology development to itself country. They indicated one example, such as material incentives to produce disappeared throughout Russia and, when Society leadership called off the experiment, the country faced industrial output reduced to 10% of what had been registered in 1914 and agricultural output reduced to such low levels as to cause widespread famine.

Why would UK encounter disastrous economy consequences if UK government did not encourage manufacturers spend money to invest to innovate automatic technology industry? According to a variety of anthropological studies, a collectivity is unable to operate efficiently with everybody giving talent workers have chance to devote whose best effort to manufacture any high technological products, e.g. human intelligence vehicle or airplane. Hence, economic incentives are needed to UK manufacturers to invest high technological automatic industry development. Because the economists predict UK will have many talent worker numbers, their number will be more than a certain number of normal effort workers, due to UK technological education level is very excellent to provide to train many young technological manufacturing students to find this kind of high technological manufacturing job. So, the high technological manufacturing job seekers will increase and it won't decrease to UK job market in the future.

Assuming that UK high technological automatic manufacturing workers who would desire only to introduce changes in the workings of the international economic order and policies of countries participating in the present economic order rather than change the order itself, what will be UK manufacturers their specific economic preferences in the future? It implies tnat either concentrate on spending more investment to automatic high technological development, e.g. human intelligence automatic high technological products or still concentrate on spending more investment to common traditional technological products.

However, UK was a developed Western country which had had strong automatic high technological development effort very long time. Otherwise, it compared to some developing countries, such as Asian China, Hong Kong, Korea etc. Asian countries their future economic growth rate will show un-surprising , different patterns, so the Asian countries has weak effort to invest high automatic technological product development, such as human intelligence technological development. The catching-up process suggests low economic growth rate in the high automatic technological product development to the Asian developing countries in the future.

Hence, the future economists predict that it views as probable successors of the Western world economic leadership if any Western country , such as UK manufacturers who prefer to invest to any high automatic technological products development , e.g. developing on human intelligence automatic technological products more than traditional common technological products development. On the one side, but it seems important to stress that two very poor countries among the challengers-China and India-are examples of countries that changed their institutions and economic policies from no or little economic freedom to more economic freedom. Because there two countries whose governments prefer to lend loans to encourage their country manufacturers prefer to invest high automatic technological products manufacturing. On the other side, attitudes toward foreign direct investment (FDI) have undergone change since the 1960 s and a large majority of less developed countries, e.g. China and India are now competing strongly among themselves and with developed market economies for direct investment from multinational companies. So, UK will face China and India high automatic technological product competitors in the future. And in fact, all countries that joined Western developed economies did that without much (if any) external inflow of public resources. It is right time that UK government needs to lend loans to encourage domestic manufacturers to invest high automatic technological products to raise whose international high technological products sale effort to win its future competitors. So, machine resources will be increased demand to o UK manufacturers if who chose to spend machine resources to innovate to manufacture any new and high technological automatic products to raise human daily life needs in the future. It means that it is right time UK manufacturers need buy much machines to prepare to manufacture many future high technological automatic products when these machine prices are low. Because the future global machine prices will possible be raised if many China and India manufacturers will also buy many machines in the future. For example, USA government had provided much financial support to assist sugar cane producers to develop their businesses. And they are dependent to a much larger extent than sugar cane producers and sugar processors in the USA on government. Without very high subsidies to renewable energy generation, they would not have survived at all. So, USA government had been the first country which could lent much financial assistance to encourage domestic renewable energy generation manufacturers to develop high technological energy manufacturing business. So, UK government needs follow USA to lend financial assistance to encourage domestic high technological automatic industry development.

Future economists also predict China and India will be competitors for future leadership in the global economy, special high technological products. China has been the media and analyst's favorite for quite some time. Quantitative projections have seemingly supported such expectation. Such as China and India had manufactured many high technological new space rockets products, ocean war large ships etc. Moreover, China has become one of the major world trade players in the early twenty-first century.

Many long-term forecasts, assuming similarly high economic growth rates in the decades ahead, predict that China will surpass the USA in terms of aggregate GDP somewhere between 2020 and 2030 or later, say between 2030 and 2050 year. The future economists conclude on the basis of these predictions that China will not only pass the USA in aggregate product (GDP), but its economy and economic policies will influence the rest of the world to a similar extent that the USA does at present.

I stressed a very important point, namely that the UK future high technological automatic product competitor China and India, namely that economies not only grow, but in the process change their structure. China and India have been industry very rapidly (the first transition) and building the physical infrastructure that accompanies industrialization changes to technology in the future. However, at a certain per capita GNP level the two countries, such as China and India will face another structural shift when which technological development will reach the mature stage in the future. China and India had been primarily historical pattern of economic development because the shift in the role of engine of growth from industry to services is to a much greater extent a qualitative shift. Both higher and different skills are required. And, even more importantly, interactions generating ideas driving the highly human-capital-intensive service economy require a much freer environment, not only in the economic area. Chinese exports have been heavily labor-intensive. This being the case, they contributed to the expansion of industrial employment, offering for the first time in the history of China a taste of (very modest) prosperity to more than 100 million new industrial workers and their families. This is the major component of the success accomplished by Chinese economic growth. Richer trade partners create room for more trade, so the Chinese should hope that intra-South trade, that is, trade between the emerging economies of Asia, the Middle East, Africa and Latin America, will open up new and growing opportunities. I presume that if Western economy , such as UK did not developed high technological automatic industry to stable their social welfare, so thoroughly slowed down their economic growth.

Will it allow China to accomplish the transition to a mature, innovation, service-sector-based market economy? It has allowed the economy to industrialize much more successfully, even if the labor shift from agriculture to industry has not yet been completed. But it is a long way off the next major test: the second high technological industry transition of the economic structure to China. Bear in mind that Russia attempted it twice and failed at both attempts.

But even, assuming that China at some point in the future does succeed in accomplishing the second transition, will it be able to supersede the USA, for example, as the main global high automatic technological innovation center if it wants to become the No.1 global high technological industry economy? Given the nature of the centralized state and its stability to collect financial resources , China's ability to increase research and development expenditure to high automatic technological products and to hire a mass of researchers, engineers, technicians and other specialists should not be doubted. This process in already taking place.

But , again, Soviet Russia already exceed the USA in the R&D/GDP ratio in the 1970s, long before the communist collapse, with no effects on its innovativeness. Inputs matter less than outputs, quantity in the innovation process mean much less than quality. The latter characteristics depends importantly on economic, civic and even political institutions. Otherwise, independent India had three options open to it in 1946s. It could pursue spontaneous economic development, with some state intervention to be sure, along the lines of basically free market capitalism; it could turn the clock back and try to recreate the rural-agricultural and handicraft based. The dominant way of thinking was Society -style priority to industrialization and , within industralization , priority to heavy industry. In other words, not textiles and clothing, which has been developing well in India since the mid- nine teen century, but production of sewing machines and , even better, production of machines the produce sewing machines.

The results were only to be expected. The heavy stress on the expansion of capital-intensive heavy industries in a very poor country quickly strained the ability of the Indian economy to generate adequate savings. Moreover, some of these industries were above the level of industrial competence of an underdeveloped economy. Thus, the amount of required resources (capital, skilled labor) was usually larger per unit of output than in the same industries in more mature, richer industries economies. In another view point, India will develop light industries, just as any other poor country with a great deal of unskilled labor, had a comparative advantage and no less importantly, an economy in which, due to their low capital/labor ratio, light industries could employ many more people, spreading prosperity more widely in a poor country. So, it explain that why China will have more effort to develop heavy high technological industry in the future. Thus, India got less economic efficiency, less employment than in a spontaneously developing economy, less ability to compete internationally in light industries suitable for an underdeveloped economy and finally got heavy industry unable to compete even on the domestic market and, therefore requiring no less heavy a dose of protection. Overall India got an underperforming economy, in particular in its relations with the rest of the world.

To conclude by comparing the performance of the traditional sectors of the Indian economy and the performance of its modern, human -capital-intensive subsector of manufacturing and skill intensive service sector. The latter both employ workers with high-and medium -high skillful level (in branches ranging from computer software and biotechnology and pharmaceutical high technological light industry). India is ahead of China in terms of the output and export of such products and services. Thus, it implies that UK ought concentrate on developing high automatic heavy high technological industry, e.g. human intelligence technological products because these industry is not better development to other many countries' strong effort , such China and India large population countries.

Consequently, future (AI) robotic technology can be applied to medical service industry, e.g. in hospital and clinic environment to let patients to live in these places to feel more comfortable. It can also be applied to manufacturing industry to assist productivity performance rasing and computer software and biotechnology and pharmaceutical high technological light industry to improve computer technological software development and invention of much new biotechnology and pharmaceutical medicines for human health.

4 Increase development in genetics, human intelligence, robotics, nanotechnology, 3D printing and biotechnology technological industry

In US future, (AI) robotic tools will assist nanotechnology, 3D printing and biotechnology technological industry development, these kinds of jobs will be needed to increase development in genetics, human intelligence, robotics, nanotechnology, 3D printing and biotechnology. For example, smart systems homes, factories, farms grids or cities will help tackle problems ranging from supply chain management to climate change. The rise of US economy growth will allow US people to monetize everything from their empty house to their car in US. These new technological products development will change US patterns of consumption, production and employment adaption are also be changed by US corporations, US government and individuals.

Why will the technological revolution be broader socio-economic, geopolitical and demographic drivers of change to influence future US social economic and consumption pattern change? Future US most occupations will also be changed. When some traditional old jobs are threatened by redundancy and other new technological jobs will grow rapidly, existing jobs are also changed in the skill sets required to do them. The debate is between some economists foresee limitless new job opportunities and foresee massive dislocation of US jobs. In fact, the reality is highly specific to future US high technological production industry, region and high technological occupation in question as well as how US production workers can be raised themselves ability to actions the upgrade level of high technological production ability from various stakeholders to manage high technological production method change.

Overall, this is a modestly positive outlook of US high technological production employment across future most high technological production industries with jobs growth expected in several sectors. However, it is also clear that this need for more talent in certain job categories is accompanied by high skills instability across all job categories. Combined together, future US net job growth and skills instability result in most US businesses with face major recruitment challenges and talent shortages, a pattern already evident in the result and set to get worse over next five years in possible.

The question is how US businesses, government and individuals will react to these new technological job changes, due to talent shortage, mass unemployment and growing inequality challenges will encounter in future US society.

The current technological revolution does not need become a race between humans and machines , but rather an opportunity for work to truly become a channel through which US people recognize their potential. So, if US traditional low manufacturing skillful workers lack talent to learn new skills to prepare to do future new technological manufacturing jobs, such as 3 D printing, robotics, nanotechnology, biotechnological high technological products manufacturing jobs. Then, it will cause increasing of unemployment rate to some not talent US low manufacturing skillful workers. So, US government or high technological product industry employers need to consider this future unemployment challenge will be caused by high technological products manufacturing changing influences. It seems high technological development will cause these low manufacturing skillful workers unemployed rising numbers as well as high manufacturing skillful workers human capital shortage global challenges will exist.

In the future, the driver of changes to influence US demographic and socio-economic growth. They may include: changing work environments and flexible working arrangements. It means new technologies are enabling workplace innovations , such as remote working, co-working spaces and teleconferencing. Rising of the middle class in Asia markets. It means the world's economic center is shifting towards the Asia developing countries.

Some economists predict that Asia will be projected to account for 66%

of the global middle class and for 59% of middle class consumption by 2030 year. In addition, climate change, natural resource will be constraints to a greener economy. It means that climate change is a major driver of innovation as organizations search for measures to help adjust to its effects. As global economic growth consumers are needed to lead to demand for natural resources and raw materials, over explanation implies higher extraction most and degradation ecosystem and these challenges will also impact US employment changes needs. All US government also needs to concern future global economic change influence. Hence , future (AI) robotic tools will assist these industries' technological development and creates more new jobs.

Artificial Intelligent

Social Military Defense Weapon

Although (AI) can influnence technological development to bring positive impact to bring beneficial welfare to provide human life. But, I also feel (AI) can bring new military to attack weak effort countries enemy from strong owning (AI) military defense wepon countries. If one day, some owning strong (AI) technological development countries' leaders who applied (AI) technology to manufacture social military defense weapon robots. Then, it will cause the third World war in possible. So, different countries' leaders need to consider (AI) invention ethic issue to keep world peace.

Nowadays, artificial intelligence (AI) is widely knowledge to be one kind of the dramatic technology. However, it is expected to continue, to have a disruptive impact on human's private and public life, so defense and security will be no exception. But how exactly will these be affected ? How will (AI) defense and security is incremental in nature?

To research why artificial intelligence (AI) has possible to be used to cause autonomous weapons by human. We need to understand these three aspects of relationship. They include cybersecurity and artificial intelligence and machine learning and autonomous weapon systems relationship between of them.

Firstly, we need to know what is the mean of artificial intelligence and cyber defense/offense? It means defense of critical networks: real time, pattern finding, anomaly seeking, it must utilize machine (AI) learning algorithms to efficiently, and instantaneously respond to potential network threats as well as it means human on or out of the loop. On the loop : it means anomaly detection: human notified, IT analysis, response. Out of the loop: it means anomaly detection: (AI) decides best method of response: quarantine, honey pot monitoring, hack-back. Thus, it is possible that (AI) can be used , such as autonomous cyber weapon.

What is artificial intelligence and autonomous weapons? Autonomous weapons mean one kind of weapon that can be selected and engaged a target, without intervention by a human operator. Are these machines artificially intelligent? I believe the answer is not, because present weapons systems are not capable of human level reasoning. But, (AI) algorithms are presently employed to process sensor data, monitor system health, take and respond to vocal commands manage data, navigate. This, future autonomous weapons systems will require stronger (AI) to be secure and operationally and cost effective. Moreover, self-aware autonomous cyber systems are crucial.

What is cybersecurity mean? It means the ability to control access to networked systems and the information they contain. It is acted to prevent , detect, recover, react. It is application objects concern people, process, technology and it's application goals are confidentiality, integrity and popular availability. Thus, what is cyber weapon mean? Walware means viruses, Trojans, zero-days, worms ransomware, spyware etc. Does it require a particular objective? E.g. military paramilitary or intelligence. Does it require physical harm? E.g. functional harm or interruption? Mental harm? Is (AI) a technological weapon that it is an object or tool? What about when it is an weapon agent?

In simplicity, (AI) can be one of scientific weapons platform. When one day, it is invented to be applied to control war planes to fly to any countries to attack enemies or it is invented to be seemed to human to replace soldiers to bring guns or any weapons go to other countries to attack. So, it is possible that future any war defense planes, (AI) technological automatic control weapon can be replaced of human soldiers or war plane pilots to control any war defense planes to go to different enemy countries to attack them easily. It is very horror matter to threaten global human's ourselves life in the future , if (AI) automatic control war defense planes or (AI) automatic control machine soldiers were invented successfully.

Hence , when (AI) can be applied to weapons platforms, it structures that launch weapons, i.e. jets, ships, vehicles. (AI) platform and weapon and software architecture components are be done one (AI) technological weapons systems. Thus, human will encounter any (AI) benefits or risks (threats) causes in the same time as soon as possible. If we can predict when (AI) weapon system will be manufactured or invented successfully. Then, we can reduce (AI) weapon systems risks , if we can threaten any (AI) scientists continue to invent any undiscovered (AI) weapons in any time to avoid the future first time (AI) weapon war occurrence in possible.

The (AI) weapon system risk means autonomy: the ability to problem solve technological war , when (AI) weapon system is manufactured successfully, the power to act, how to damage the (AI) weapon system. The power to chance to stop (AI) weapon system manufacturing processes, ability to create a new goals, how to change the (AI) weapon system inventors' or scientists' minds to avoid to apply (AI) tools to achieve attack goals to change to another positive goal. Due to human can't know a prior what an autonomous (AI) weapon system will do.

Although, human is known what (AI) is , but human is also known when (AI) scientists whose emergent behaviors will do to change to do any negative behaviors from positive behaviors. Whatever (AI) weapon system design we use, there will be cybersecurity, problems arising from computation design/complexity. Due to any one (AI) scientist can manipulate the system to act against itself, or who can utilize traditional " cyber weapons" against the (AI) weapon system, or who can manipulate the system to lie to humans, but also due to complexity, there is no way to know if it is lying or not or bounded rationality : satisficing.

Finally, the most serious (AI) technological invention risks are human is unknown these aspects of (AI) absolutely: They are not simple automatic systems, learning reasoning, communication of " self-aware" systems. Thus, human will face (AI) technological invention risks or threats. We need to find any methods to avoid (AI) weapon system is manufactured successfully to avoid (AI) technological war can occur in future anyone day.

1 (AI) system immoral intention

Why (AI) system can be invented to damage our society ? IS it possible to achieve this (AI) damage system successfully? ON (AI) attribution hand, it can be applied to cars, aircraft, which are subject to regulation designed to protect the public from harm and ensure fairness in economic competition. Thus, (AI) safety issue is important to scientists to consider.

IN general, the approach to regulation of (AI)-enabled products protect public safety issue should be informed by assessment of the aspects of risk that the addition of (AI) way reduce any respects of risk that it may increase. Also, where regulatory responses to the addition of (AI) threaten to increase the cost of compliance, or slow the development or adoption of beneficial innovations, policymakers should consider how those responses could be adjusted to lower costs and barriers to innovation without adversely impacting safety or market fairness.

For example, regulatory challenges that (AI) enabled present are found in the cases of automated vehicles. (AI)s, such as self-driving cars and (AI)-equipped unmanned aircraft systems. IN the long run, self-driving cars will likely save many lives by reducing driver error and increasing personal mobility, it will offer many economic benefits. Thus, public safety must be protected as these technologies are tested and begin to mature. Creating safe spaces and test beds for experimentation , and working with industry and civil society to evolve performance based regulations that will enable more uses as evidence of safe operation accumulates. Thus, it implies that any scientists can also invent (AI) system to control weapon defense planes or (AI) automatic machine human to do any soldier's behaviors to attack to any countries easily, instead of none driver automatic control vehicle invention. Thus, (AI) system can be applied to harm to human or achieve to damage our society aim by ourselves in possible.

The rapid growth of (AI) has dramatically increased the need for people with relevant skills to support and advance the field. AN (AI) –enables would demand a data literate citizenry that is able to read, use, interpret and communicate about data and participate in policy debates about matters affected by (AI). Thus, if (AI) technology is applied to assist human's social development and raising life enjoyment or benefits. It will bring positive impact to influence human's future life. Otherwise, if (AI) technology is unsafe to be applied to threaten human's society. It will bring negative impact to influence human's future life. Thus, (AI) scientists need to consider how to apply (AI) technology.

As (AI) technologies move toward deployment, technical expects, policy analysts and ethicists have raised concerns about unintended, consequences of adoption. Use one (AI) to make consequential decisions about people, often replacing decisions made by human –driven bureaucratic processes, leads to concerns about how to ensure justice, fairness, and accountability, the same concerns of human's safety issue. Thus,)AI) expects have cautioned that there are challenges in trying to understand and predict the behaviors of advanced (AI) systems.

Use of (AI) to control physical-world equipment leads to concerns about safety, especially as systems are exposed to the full complexity of human environment. A major challenge in (AI) safety is building systems that can safety transition from the closed world of the laboratory into the outside open world, when unpredictable things can happen. Adapting to unforeseen situations are difficult necessary for safe operation. Experience in building other types of safety artificial systems and, such as aircraft, power plants, bridges and vehicles has much to teach (AI) practitioners about verification and validation, how to build a safety case for a technology, how to manage risks, and how to communicate with stakeholders about risk. The risk means the harm of human's safety of (AI) damage system control machine invention. Thus, any (AI) scientists need consider moral responsibility when who decide to invent what kind of (AI) system machine to aim to bring human's benefits or attribute to human's welfare intention.

Thus, (AI) products safe invention matter will need any scientists' considerations. Because , if (AI) any products are unsafe or harm human's invention in the manufacturing process, it will bring any human's life danger when the (AI) system damage tools are invented successfully and are provided weapons to humans to use to attack other countries easily. It will cause future global human (AI) technological war occurrence.

I shall recommend the solution is necessary of ethical training for (AI) practitioners and students. Ideally, every student learning (AI) , computer science, or data science would be exposed to curriculum and discussion on related ethics and security topics. However, ethics alone is not sufficient. Ethics can help practitioners understand their responsibilities to all stakeholders, but ethical training should be methods for deciding good intentions into practice by doing the technical work needed to prevent unacceptable or immoral (AI) invention outcomes.

Hence, global human needs to concern (AI) weapon system invention security issue. Nowadays, (AI) has important application is increasing role for both defensive and offensive cyber measures. Currently, designing and operating secure systems requires significant time and attention from experts.

Challenges issues are raised by the potential use of (AI) in weapon systems. The United States has incorporated autonomy in certain weapon systems for decades, allowing for greater precision in the use of weapons and safer, more humane military operations. Nonetheless, direct human control of weapon systems involves some risks and can raise legal and ethical questions concern (AI) manufacturing process intention.

The key to incorporating autonomous and semi-autonomous weapon system into American defense planning is to ensure that U.S. Government entities are always acting in accordance with international humanitarian law, taking appropriate steps to control , to develop standards related to the development and use of such weapon systems. The United States has activity participated in ongoing international discussion on Lethal autonomous weapon systems and anticipates continued robust international discussion of those potential weapons systems. Thus, (AI) scientists have responsibilities to manage the potential to be a major driver of economic growth and social progress only, their (AI) intentions are not the global dominance aims absolutely, if (AI) product industry , civil society, government and the public work together to support (AI) positive development of the technology with thoughtful attention to its potential and to managing its invention threat risks to avoid (AI) products to manufacture to be used weapon tools.

Finally, I recommend that as the technology of (AI) continues to develop, practitioners must ensure that (AI) enables systems are governable, that what their inventions need to be openness to let public to know clearly and understandable; that they can work effectively with people and that their operation will remain consistent with human values and aspirations. Researchers and practitioners have increased their attention to these challenges , and should continue to focus on their future any (AI) inventions.

Hence, (AI) safe system ought to be applied to solve the biggest challenges that society faces, such as mobility for the elderly and those with disabilities, smart buildings may save energy and reduce carbon emissions, precision medicine may extend life and increase quality of life, smarter government may solve citizens more quickly and precisely., better protect those at any immoral invention risk and save money.

Moreover, (AI) enhanced education may help teachers give every child on education that opens doors to a secure and fulfilling life. Thus, these are the future human's potential benefits if the (AI) technology is developed to its benefits and scientists ought avoid to manufacture (AI) tools to cause weapon risks and challenges.

Consequently, the main point is that how experts invent (AI) systems. (AI) systems ought not be advanced weapon systems, it doesn't seem to be thought similar human soldiers mind and behaviors. (AI) system ought be systems that think like humans. (e.g. cognitive architectures and neural networks), systems that act like humans (e.g. pass the test via natural language process, knowledge representation, automated reasoning, and learning), systems that think rationally , e.g. logic solvers, inference and optimization and systems that act rationally e.g. intelligence software agents and embodies robots that achieve goals via perception, planning reasoning, learning , communicating, decision-making and acting function.

In conclusion, it is horror (AI) scientists will invent (AI) systems to be owned human's (soldier's) mind and attack strategic behavior to attack other countries easily, who must need to consider (AI) system ought be invented to own scientists' creating mind and non manual assistance functions for positive attribution to human's society. I expect that (AI) system can only be invented to create human's welfare in our future.

2 (AI) soldier weapon ethical, social and
economic negative impact

In the future, how human can avoid (AI) technological ethical, social and economic negative impact. Scientists need to concern these questions: how to develop of a good (AI) society, how the role and responsibility of the government, the private sector, and the reserch community(including education), in pursuing such a development, whether how the recommendation to support , such a (AI) system development may be in need of improvement.

However, none appers to deliver a comprehensive explicit vision of the role that (AI) system should play in mature information societies. Thus, (AI) 's potential contribution to social good shoud include an in-depth plan for linking in a comprehensive socio-political design questions of responsibility of the different stakeholders, of cooperation between them and of sharable values to understand of a good (AI) positive impact society, not a bad (AI) negative impact society.

Thus, the notion of mature information societies is introduced to stree the importance of addressing the current ethical challenges that (AI) poses in a comprehensive fashion.

It seems (AI) wil invention will be human's moral societal consideration issue. It concerns our (AI) scientists' moral issue, how who invent (AI) system to apply to which kind aspects. IF (AI) system was one direction on war weapon tools to similar to soldier's personal mind or attacking behavior. Then, it will bring poor social safety and poor economy growth our world, due to (AI) scientists' moral is low level.

Thus, the developed country US (AI) technological leader needs to focuse on the impacts of (AI)-driven customatin on the US job market and economy. It represents three specific policy responses to the perceived impact of (AI) on the US economy. They include these three aspects such as: How to invest in and develop (AI) for its many benefits, how to educate and train Americans for the jobs of the future and how to aid workers in the transition and empower workers to ensure broadly shared growth.

The future of (AI) influenced cyber conflicts need more than just the application of current and past solutions in order to ensure security and stability of societies, and avoid risks of escalation. To achieve this end, efforts to regulate cyber conflicts require an in-depth understanding of this new phenomenon, identify the changes brought about by cyber conflicts and the information revoluation, and defines a set of shared values that will guide the stakeholders operating to avoid the international (AI) war occurrence. This becomes clear when considering for example, cyber deterrence. Deploying conventional (cold war) strategies to deter (AI)-influenced cyber conflicts proves highly problematic and the urgent need to foster and coordinate new solutions able to account for the any kinds of conflicts of the cyber demain and of mature information societies to avoid (AI) technological war occurrence in the future.

We hope that in the on-going international conversations and reviews, the US government with further specify how " (AI) system invention law" fit into their vision of the future of society in this case the future of (AI) technological war and conflicts. Hence, (AI) scientists need to concern ethical issues related to (AI), like fairness, accountability and social justice can be addressed through increasing needs. Such as: how the creation of a new body focused on robotics and related (AI) system development to avoid to intent to apply weapon tools to provide advice on the policy, legl and consumer protection issues arising in these fields should be considered.

How to achieve ethical training of (AI) staff and ethical education of the public is certainly important responsibility for (AI) tools ethical behavior and design to the private sector and the citizens : of unique challenges that (AI) brings to society in terms in fairness, social equity and accountability are addresses. Thus, the development of the (AI) technology and defining good (AI) remains problematic. In particular, the US government's innovation driven approach to defining the potential, positive impact of

(AI) shows that more could be done to ensure that the opportunities and advantages brought about by (AI) are shared by all society.

An initial on Robotics, based upon the ethical framework and guiding principles is proposed. It should be complementary to legislaton and comprise ethical codes of conduct for Robotics researchers and designers, codes for research ethics committees as well as licenses (rights and duties) for designers and users. Thus, (AI) robotics invention of safety issues is very important considertion to any (AI) inventions or researchers. Every country's government ought have legal guiding to control their robotics' manufacturing intention. If their robotics (AI) is applied to seem to be soldiers to attack other countries to threaten their people's safety. Then, those (AI) inventors or researchers need to be punished by law.

In conclusion, I believe (AI) technology will be applied to weapon, when it's technological development is nearly mature to able to learn human's mind to do any behavior. During (AI) technology reachs thie mature stage, I predict the (AI) weapon tool , e.g. (AI) soldiers will have chance to be caused. This (AI) invention mature stage has these characteristics such as:

When (AI) invetion reachs this mature stage, computers and robots will develop conscious, intelligent, personified minds. Further, information technology devices and (AI) systems will be implanted into humans, enhancing, psychological and behavioral abilities and allowing for direct communication with artificial intelligent minds. There will be both artificial intelligence (AI) and intelligence amplification (AI) in the relatively near future stage.

During the (AI) invention reachs this mature stage, these will be an ongoing mulit-faceted integration of information technologies and human life. Humans and information technology will cooperate. Humans will increasingly immerse their lives and minds in (AI) systems of technological intelligence and virtual reality. The distinction between humanity and technology will increasingly close dependence.

During the (AI) invention mature stage reachs that the environment will be infused with information technology, becoming animated, communicative and more intelligent. The destinction between the artificial and the natural will increasing close dependence.

During the (AI) invention mature stage will expand through virtual reality, simulated and virtual reality will increasingly into normal reality, e.g. the (AI) weapons is virtual reality to seem to be soldier weapon.

Finally, during the (AI) invention mature stage is as the global expression of the evolving human-technology integration a " world brain" and " world mind" will emerge on the earth. This psychophysical (AI) weapon system will enhance and enrich the capacities of both individual and collective cogniton. This (AI) weapon system is a potential starting point toward the evolution of a cosmic brain and cosmic mind.

Thus, it is possible that the workship raw data was a unique way in which (AI) could be weaponized to cause war, during the (AI) invention stage reachs the invention mature stage. However, (AI) weapon manufacturing factory will be built possibly. In the future, how will we defins and locate (AI) weapon factories. Especially, as these factories are no longer solely buildings , but a mil of virtual and substantially different facilities, particularly as it shifts from a physical assemly and development model to a distributed and flexible network. Needing minimal raw materials to develop (AI) weapons, the phsysical location of their (AI) factories could be anywhere and their identification from the outside, nearly impossible. Given the expanding uses for intelligent and super-intelligent (AI). How will we tell the different form a location that is manufacturing (AI) for the creation of weapons versus creating (AI) for an innovative new gaming platform?

In conclusion, human needs to consider every (AI) scientist's personal ethical or moral mind and research intention and (AI) system invention of (AI) weapon factories cause. During (AI) invention reachs the mature stage if human expects to avoid (AI) technological war occurrence in future one day. The technological development on autonomous military robots, ideally among relevant social groups and actors including human-rights, activists, researchers developers, engineers, philosophers, policy-makers, military authorities, lawyers, journalists and the publis need to consider when human has effort to invent autonomous military robots successfully in the future one day. Finally, some ambitious countries or dominant global countries must like to apply (AI) autonomous military robots to be machine soldiers more than human soldiers if (AI) technology had reached the mature stage. So, future (AI) autonomous military robots will be the next choice of weapon to follow nuclear weapon. If civilians were used as a human (AI) soldiers, the weapon simply ignored them and targeted anyway. This scenario highlighted the dangers of proliferation and quick replication of autonomous weapons. Unlike nuclear weapon, a piece of code for (AI) artificial intelligent soldier could be obtained on the black market and replicated at little cost and the hardware for this type of weapon doesn't require costly or hard to obtain components and materials. Thus, (AI) artificial intelligent soldiers can be manufactured many at cheaper cost. Otherwise, manufacturing one nuclear bomb weapon will spend too much cost. Hence , it is possible that (AI) artificial intelligent soldier will be future new technological weapon to follow nuclear bomb weapon. Hence, any country government needs to legislate to control any (AI) scientists' inventions whether they are attributed benefits or welfares to human or damage human's safety.

(AI) assist future computer
industry new gender innovation development

Why China's computer manufacturing and product development industry will be global leader to compete US computer dominant market. The reason is because that China will have possible to dominate global computer industry development if it can invent new (AI) learning tool to assist global computer systems to raise more efficient performance effort.
Nowadays, China's computer industry is the largetest hardware producer production and experts is dominated by Taiwanese firms. It is also the second largest personal computer (pc) market and domestic pc companies are top three sellers in global computer manufacturing and product development market. Forx example, Lenovo buys BM pc business in 2004 year. It implies US, IBM pc manufacturing leader can not dominate global computer market in possible in the future.
Reed Electronic Research, Yearbook Of World Electronic Data (2003) indicated that the leading computer producing countries of hardware production in US $millions and share share of total gogal production: The world region US was the global rank number one. In 1995 year, US had US $76,284 value, market value 26.5%. Then in 2000 year, US had increased up to US $ 90, 430 value, market share 24%. Till to 2003 year, US had fallen down to US $ 69,102 value, market share 21.7%. However, US hardware production was still the global rank number one , although its hardware production value had been falling down. But, the following second rank country, Japan and the third rank country, Singapore and the fourth rank country, Taiwan and the fifth rank county China which hardware production value could not exceed US till to 2003 year. However, although China had the lowest hardware production value US $5,600 to compare to among of these countries in 1995 year, but China had increased the value to US $65,000 and market share to 20.5%. Otherwise, Japan, Singapore and Taiwan value and market share had surprisingly fallen down below than

China value in 2003 year. Thus, it seemed that China will be a potential country to compete US hardware production industry after 2003 year.

Reed Electronic Research, Year book Of World Electronic Data (2003) also showed that these computer companies of China had these % of market share : Beijing Founder had 9.9%, Tsinghua Tongtang had 7.8%, dell had 7.2 % , IBM had 5.1% , HP had 4.8% of market share. Thus, it also seemed that China some computer companies will have impotant large market share percentage in global pc sale market. In the future, global hardware production and pc sale industry. China and Taiwan both countries will be one pc manufacturing and design and sale partner. The reason is that China and Taiwan had been the number one rank of markers of notebook pcs, motherboards, scanners, keyboards, add-on card optical drives, monitors and some network equipment etc. pc (personal computer) relative computer function products. It seems that these both countries had co-operated to research any computer relative products to sell to global computer market. They are also the original design manufacturers (DDMS) develop and manufacture over half the world's notebook pcs as well as their customers include all major branded pc vendors (OEMS).

Taiwan Minstry Of Economic Affairs (2003) indicated Taiwan's top notebook ODMS include: In 2003 year volume (thousands) Quanta had $8,500 sale volume thousands , for example, Quanta major OEM partners include Gateway, Dell, HP, IBM, Apple , Sharp, Sony, Fujitsu-Siemens (F/S). Compal had $6,000 sale volume (thousands) , Compal major OEM partners include Dell, HP, F/S, Toshiba, Acer. Thus, it also implied Taiwan had many small size and non famous brand of computer companies which choose to co-operate to be partners with some global large size and famous brand of computer companies to raise competitive effort in global computer market, such as Dell, IBM, HP, Gatway, Apple etc.

Thus, the future trend of computer new product manufacturing development will shift from US to Taiwan and SE Asia, then to China. However, what kind of knowledge work factors will be needed to China and Taiwan . In general, notebook manufacturing stages will include: The first process is design stage, it includes concept design, such as analyze need, create concept and set brand image as well as product planning, such as business case, specifications, industrial design and sourcing strategy. The second process is development stage, it includes design review steps, such as design review, such as mock-ups, electrical test as well as prototype build, such as commercial samples, integrated system test as well as pilot production, such as production process design, pilot. Final process is production stage, it includes mass production, such as ramp-up, volume production, production testing and global distribution as well as sustaining support, such as speed bump, component replacement, technical support and warranty support. Thus, I believe that China and Taiwan must own thee knowledge work skillful of computer design and development professionals who can assist these two countries how to innovate their future computer development to change global traditional computer model to be renew and innovate computer model in the future.

Due to computer industry's stages of development and manufacturing are closely linked , need manufacturability , testing of sample products, concept design and product planning stay together in lead markets and branded vendors, design and development can be separated organizationally and geographically. Thus, China and Taiwan choose to co-operate to exchange their different skill, such as either China has own more concept design and product planning skill or more development skill or more production skill. Then, China will choose either one of the most beneficial comparative advantage among of them. To bring this one of the most beneficial co-operative advantage to attract Taiwan to choose either one of the beneficial comparative advantage of skill, such as either design or development or producton to already co-operate to compete the Western developed country US together.

Thus, US won't be the global computer industry development leader if both US country famous and large employee number computer companies, such as IBM and Apple which choose to outsource their pc design and development and production skill to China and Taiwan both countries to help them to develop global computer design and development and production skill to be upgraded. Thus, I feel these both countries will plan how to co-operate to compete US to win the global computer industry leader position in the future.

When China invented its (AI) learning system success. Why does it influence global computer industry market change? For example, in the future, instead of global computer manufacters need to consider the design, development and production processes, who also need to consider what factors can influence consumers' laptop purchases. Because any consumers have much different computer model and brand to choose to make final decision to buy any computers. If the computer manufacturer can predict what factors will be whose weakness(es) to influence global computer consumers to change whose mind or attitude to choose to buy other brands of computers, then it won't lose its many old computer customer numbers and reduces it market share in global computer market share.

Nowadays, in general computer has three kinds to provide to global consumers to choose to buy , such as laptop, notebook computers, desktops. it seems that laptop and notebook computers and desktops will have different factors to influence any consumers to choose to buy any brand of computer products. Thus, computer indsutry can divide three consumer groups, such as (stayers, satisfied switchers and dissatisfied switchers) of a computer company with respect to the factors influencing consumers' laptops or notebook computers or desktops purchases. However, I feel the factors can include such as core technicl features, post purchase services, prices and payment conditions, peripheral specification, physical appearance, value added features and connectivity and mobility seven main factors that are influencing consumers' laptop or notebook computer or desktop purchases in global computer industry market.

Ganesh et al., (2000) indicates the customer base of a company consists of three groups of consumers: stayers, satisfied switchers and dissatisfied switchers. Therefore, the consumers in this study replied to the question about whether the current brand that who were using was their first laptop brand or whether who had switched from a previous laptop brand. As a following question, consumers who had switched were asked to state the reason of why who switched from a previous laptop brand brand to their current brand. The options include overall dissatisfaction from the previous laptop brand and reasons other than dissatisfaction. Thus, computer companies need to know what factors influence either whose prior computer customers why who don't choose repeat to buy its any computer products or whose new potential computer customers why who don't choose to buy its any computer products in the first time choice. Thus, future computer manufacturers need to consider intangible salespeople service attitude or performance, such as salespeople current purchase and post purchase service, e.g. technical repair, model function explanation how to use the computer, instead of tangible product performance, e.g. computer appearance design , function , mobility and internet and document download speed connectivity function. Because salespeople and technicians' service performance can be represented to the computer image. If they can provide excellent service to let computer buyers to feel satisfactory, then they can help their computer company employer to build good image. So, staff service performance will be one important factor to influence computer consumers to make the final decision to choose to buy the brand of computer products more easily. Even, one famous brand computer company, such as IBM, Apple, Gateway, these any one of famous brand computer company must not attract any new (the first time) or repeat computer buyers to choose to buy their any kind of computer products , such as laptop, desktop or notebook more easily due to their famous brand. Althoug, these famous computer companies had built good image to let consumers have more confidence to buy any kind of their computer products. But, if these famous computer companies' salepeople or repair technicians can not provide excellent customer service or performance to satisfy their computer buyers' service need, e.g. explaining how to use the new computer, repair post purchase service etc. I believe these famous brands of computer consumers will not have more desire to prefer to chose to buy any one of these famous computer brand's products. Otherwise, if the other less famous computer companies' any kind of laptop, desktop or notebook sale price is higher than the famous brand of computer companies' products sale price, but their salepeople or technicians can provide more excellent service attitude or performance to satisfy their consumers' needs. It is possible that the new or first time computer buyers or repeat computer buyers will still choose to buy their computers. So, the famous or less famous computer brand is not one important factor to influence the computer buyer to decide either to buy the computer or not buy the computer. Otherwise, computer company's salepeople and repair technician whose service performance or attitude will be one important intangible factors to influence any first time (new) or repeat computer consumers to choose to buy any famous or less famous brand of computer company's product, instead of the tangible computer design appearance and reliable function and convenient mobility and long term durability etc. factors influences.

Thus, China has possible to influence global office and home computer comsumers to choose to buy its any brands of computers to use if it can invent (AI) learning systems to assist global computers to raise their performance efficiency. So, it will influence global computer consumers to choose its country's any brands of computers to buy to use, due to themselves new (AI) learning computers can help office and home computer users to raise efficiency and provide the excellent productive performance to them more than the traditional computers.

1 Can culture factor influence the (AI) computer consumer choice?

When China's (AI) computer learning system has developed in success. Then, it will possible to influence global consumers' traditional computer applying culture to change to new innovation (AI) learning computer applying culture. It means that China will dominate global computer consumer choice to be trended to choose to buy China's any computer brands' produducts , due to it 's (AI) technology can be invented to apply to traditional computers in order to raise their efficiency and reduce office staffs' workload and provide excellent performance to serve office or home (AI) learning computer users.

Durmza and Zengin, (2011:53) indicted marketers closely interested in this issue to know the family which changed and renewed in course in time. It provides an advantage for a marketer to know the family structure and its consumption characteristics. Nowadays, consumer behavior is influenced not only by consumer personalities and motivation, but also by the relationships within families. Family is a social group and it can be considered a crucial place in th perception of marketing (Durmaz, Yakup, CELLK, Mucahit and ORUC, Reyhan, (2011).

The consumer buying behaviors examined through an empirical study. Then, it brings this question: Whether cultural factors will influnece the computer consumer choice. Choice and include computer brand choice, computer price choice, computer model choice, computer design choice, laptop or desktop or notebook product choice, new or second-hand old computer choice, the computer of manufacturing country choice, computer package choice etc. So, any consumer will consider to choose any one of these to decide to buy which kind of computer.

Every country computer consumers had different culture to influence their computer shopping choice. I feel culture can be explained how to influence to computer shopping such as: How do the country computer consumers buy and use their computer products habitually ? How do the country computer consumers react to th computer price changes, attractive advertising methods to satisfy whose needs and computer company store interiors? What underlying mechanisms operate to produce any one of the country computer consumers' responses? If computer marketers have answers to such these questions, who can make better managerial decisions how to adopt which computer target country (countries) consumers' culture.

Consumer behavior deals with many other issues, for instance (Priest, Carter and Statt, 2013: 19). How do we get information about products? How do we assess alternative products? How do different people choose or use different products? How do we decide on value for money ? How much risk do we take with what products? Who influences our buying decisions and our use of the product? How are brand loyalties formed and changed? For computer industry, it means that how computer consumers get information about computer products, how computer consumers assess alternative notebook, desktop, laptop computer products, how different age, country, culture, sex, student or working people or retired people computer consumers choose or use different kind of computer products, such as notebook, desktop, laptop computer products, how much risk computer consumers take with notebook, desktop, laptop computer products, the computer consumers' buying decisons and their use of the desktop or notebook or laptop computer products will be influenced by whom, e.g. family, friends, teacher, employer, computer salepeople, advertisement marketer etc. , computer company brands how are formed and changed by whom, e.g. computer consumers, computer company competitors, marketers, different countries' culture etc.

Durmaz and Jablonski, (2012:56) also explained culture is the essential character of a society that distinguishes it from other cultural groups. The underlying elements of every culture are the values, language, myths, customs, laws and the artifacts or products that are transmitted from one generation to the next (Lamb, Hair and Deniel, 2011: 371). Culture is the most fundamental determinant of a person's wants and behavior. Whereas, lower creatives are governed by instinct, human behavior is largely learned. The child growing up in a society leans a basic set of values, perceptions, preferences and behaviors through a process of socialization involving the family and other social roles. So, I feel different country have different culture to influence as well as different country computer consumers who have different computer purchase and consume habitually. So, computer manufacturers ought focus on manufacturing the unique need and characteristics to satisfy any country's consumers' needs.

What is my idea about future global computer competition and factors influence computer consumer behavior ?

In conclusion, future computer industry development will trend that computer manufacturers need to consider every country's computer comsumer culture. Because every country computer consumers who will have different computer consumption habitually if who can predict what the country most computer consumers culture, then they can have more confidence to sell their computers to different country markets. Moreover, US computer manufacturers need to consider China and Taiwan computer manufacturing technology because it is possible that these both countries will be its main competitor among different computer manufacuring countries. Because thess both countries will cooperate to research new model of different computers to attract global computer consumers to choose to buy their new model of computer products in the future. Finally, computer manufacturers need to consider salepspeople and repair technicians service performance because computer consumers will consider intangible service performance , instead of tangible computer quality and price and style etc. factors . The main reason is that any computer have chance to be needed to repair and salespeople' skill will influence the computer consumer to make final decision to choose to buy the brand of computer. Thus, these factors will influence global computer development and trend in the future.

Artificial Intelligent Robot: Technology change traditional production of factor model

1 What is mean of (AI) Technological innovation production of factor ?

Can (AI) robot technological learning system change future traditional production of factors model: land, human, equipment and capital to any organizations in order to replace these production of factors and assist organizational development efficiently and effectively?System may be physical , like the solar system or an ecological system or which may be simply behavioral, like an organization. For example, a national economy may be a system, markets are systems, firms and factories are systems. Even, families and individuals are economic systems. The economy of the largest systems, national economy, may be called macroeconomy, which deals in terms of national aggregates for output, income, productivity. The economics of small systems, which are their parts or subsystems may be called microeconomics.

This is traditional production of factor model. for example, a system transforms inputs into outputs. An economic system is such a process. For example, factories are as systems take in raw materials, services etc. and change them into products for sale, i.e. output and consume them, thereby transforming them into rubbish, incidential is bad output. Also, countries consume their actural resources to enhance their standard of living and change them into waste products. If the system in question is national economy, some of the subsystems are might consider to be: the government, the firms, the consumers, the natural resources which it has at its disposal. Each subsystem is itself composed of subsystem of a lower order, such as a firm and each of these can be decomposed into further subsystems, depending on the purpose of the analysis. " All subsystems" interact need have individual characteistics, i.e. they are synergistic if they expected to raise producivity or efficiency or effectively. So, it needs high technological assistance to raise whose ability in economic view.

However, an economic system must continually adapt and restructure to meet the challenges of a changing economic environment if it is to prosper. For example, a firm must respond to its environment in the form of it customers' needs threats from its competitors, government regulations etc. Nowadays, technological innovation process and the nature of social economic and social changes which is occurring as the same time. So, organizations need to have strategic management to raise technological innovation to achieve raising productivity and efficiency aim. In the futue, (AI) robots will be possible one kind of new production of factor to assist organizational development and raise manufacturing efficiency and staffs' working performance in every team.

2 How does (AI) technological innovation occur in economic process?

What is economic process? It consists of the production and consumption of products and services by human. It is a process devised by human for own benefit pupose only. In the past, human lack advanced technological invention, e.g. family society required a much greater degree of organizational skill than hunting and gathering, it seems farming society does not need to achieve efficiency or productivity aim, because it is not industralized manufacturing society. Nowadays, the investment of resourcs is required for manufacturing processes for factories. The manufacturing stage thus needs machines, but it extends the economic process into the processing of manufacturing things, such as food. So, the knowledge and skills to do this are much more specialized again than farmer's or hunter's. So, technological innovation is needed to raise efficient productivity in factories, e.g. the increasing skills of manufacturing and the use of more intensive energy resources, such as coal and oil, gas, even solar energy either resources are from the sun or resources are from earth, e.g. fuels , heat, light, sound utilitiesm liquids , gases,solids. So, technological innovation is important to influence our economic development in our societies.

Economists usually classify what who call future of production into land, labor and capital. Why technological innovation is one another factor of production. For example, the economic process indicates that the first step is resources from the earth, e.g. solar energy supplies to earth to satisfy human needs. In the economic process, it needs these both supplies, driving force of transformation energy supply and captalyst , such as skills, knowledge, organization, creativity, creative participation in consumptions supply. Then, manufacturers shall change these both supplies to production and distribution of ordered materials and utilities in the economic process. Finally, it will provide to human consumption and human spent resources returned to earth in the final step. Another example of the elements of the economic process: the input is driving force of transformation stage of energy sources, e.g. sunlight firewood, oil and gas, coal , nuclear and household and industrial waste. Next is the economic process stage: facilitators, it includes tangible facilitator includes skills, knowledge, organization, creativity, e.g. language, science, technology, industry, machine, politics, law and order, defence, strategic plan, information systems, administration, tangible facilitator includes incentive system, e.g. money, banking, insurance, shares, private or public organizations, markets, land area. Finally, is the product of innovation stage, it includes utilities , such as electricity , heat, light, sound, motive power as well as ordered materials (products) , such as bread, meat, mine, shoes, clothes, houses, television, roads (public goods) etc. In future, (AI) robotic development will be possible participate to new economic process in order to raise global efficiency and performance for every businesses.

3 How (AI) robotic innovation information factor influences the product successful sale

What is the role of (AI) robotic innovation information (big data gathering method) ? Any markets requires product or service suppliers rationally act on the basic such information. But what who can't know in advance is how all the other participants are going to behave. The market clearing price would already be known. There would in fact be agreed prices and which everything could be exchanged, and there would be no market system at all. And so who come to market to settle the price/quantity relationship. The theory is that which will arrive at a single price and quantity which reflect supply and demand. However, the number of interactions or pieces of information to be transmitted doubles with every new participants. However, the requirement for information is not limited to the particular market in question. A compromise between the number of people needed to make more nearly " perfect" in the economic sense, and the quantity of information needed to allow it to arrive at a unique price/ quantity relationship. The concept of degrees of freedom is widely used in different technological forms, e.g. engineering industry, the equipment is used by manufacturers to make pencils will be worn out to some extent in the process, and this forms an energy path straight to earth from the market in which the equipment was bought. Similarly wear and tear on the equipment used to make the intermediates and the raw materials will also form direct paths to earth from the markets in which were bought. It seems technological innovation factor of production can bring the pencil stationery product innovation when the new pencil stationey product is produced the more excellent quality by the new machines innovation.

In an economic system which is working " perfectly" according to the definitations, output is therefore a function of available energy and the technological skills to apply it to conversion of inputs into materials and utilities . In a market economy, given the availability of inputs of energy and materials, and the necessary information, the only factor which can bring this about in the long term is a change in the energy efficiency of its conversion process, i.e. the energy consumed unit of output of the same total production. This depends in the application of skills and design, that is technology factor of production.

Why (AI) big data gathering information can influence product sale ability. For example, the commodity is technologically complex like a computer, an aircraft or even a refrigetator. One is buying not just the piece of equipment, but also its specification because few people would understand the parts of the machine, let alone be able to judge their quality. Furthermore, one is also buying the future performance of the machine in operationm , its fuel consumptionm reliability, service costs, length of life, resistance to obsolescence etc. Probably the only guarantee that any information obtained on these points is valid is the reputation of the manufacturer. Purchasers estimate chose chances of surviving the guarante period. Brand names are a way of simplisfying information flows. Such problems of defining the commodity and so handling the information necessary to arrive at a stable price, are magnified when counterfeit products, such as are flooding on to the market at present, find their way into markets for genuine products. Buyers will be unable to distinguish unless who are experts and sometimes that may need chemical analysis or destructive testing. This is a recent phenomenon to buy technological products.

4 Why does (AI) big data gathering information technology influence the real market system change ?

Can (AI) big data gathering information technology be one kind of production of factor to influence the real market system change to be more fast speed of market information communication in global industries? The market system in the real world includes: the first is manual work (manpower) element, in effect the provision of an elementary utility for consumption in a conversion process. If labors are not providing manpower, who become unemployed. Unemployment is not simply leaving a resource at a particular time, it is an injustice and a burden one the very real society which economics is supposed to help. Moreover, the unemployed can't spend the money who don't earn, and so buyers are reward from the economic process. The second is catalytic skills, knowledge, organization and creativity element which applied to the conversion processes which turn raw materials into products and utilities for consumption. In this case, who are as varied as the individuals that make up mankind, their accumulated knowledge, their capability of organising themselves to achieve their ends and not least their creativity,the ability to generate entirely new catalytic effects. Finally, is the incentive element, which is the prospect of participating in comsumption of the products of the economic process. The incentive system is cash for current or future exchange for products and utilities. The incentive system must be within the control of the social system of which it is a part. It can only be addressed by society as the whole system. However, for the individual and the firm too the creative must by definition come from outside and it is also depend on the rest of society.

What kinds of product can link between markets to increase speed of market information communication when global industries choose to apply (AI) big data gathering information technology to gather global competitors' product and client and price etc. business data. However, there are products which are linked in a different way by associated use. For products anyone who buys a vehicle must also be prepared to buy its fuel, tyres etc. A decision to buy the vehicle therefore automatically generates subsequent expenditure in the other markets. These markets are not so much competitors for buyers' money as complementary to each other. Sale in one must lead to sales in the other. This technological products have the same point, it is that which are needed to attempt to innovate their quality to raise their competitive ability to win their competitors. It seems that (AI) big data gathering information technological innovation can be a factor production to these different brands of vehicles and which related link products. Much the same occurs in technological industry. A company may feel that it is wise to buy related pieces of equipment from the same manufacturer, especially if they have to be connected in some way, whatever the price, within reason.

5 Can (AI) big data gathering timing of information influence real marketing system?

The analysis has shown that two sorts of (AI)big data gathering information are essential of the market is to reach " equilibrium" values of price and quantity: information concerns on the economic environment, which participants can obtain before the market opens; and information about the process of bargaining displayed, which can only be made available as the bargaining proceeds. However, buyers and sellers happen next. If the information acts as a reference point, it can only be a historical one. This is particularly so where markets operate continuously. There is always a lapse between the conclusion of deals and their display, so that new deals are always influenced to be not update information , in the absence of the most recent data, if dealing is busy. So, timing of information ought to be kept the most update to let buyers can have more confidence to make final choice to buy the broad of products. The quality of information which had to passed during bargaining in order to achieve on a unique price/quantity relationship increased rapidly with the number of participants because of the need of to allow everyone of buyers to interact with all the others.

It follows therefore, that as the number of participants becomes very large, the necessary information flows become much larger still and the time needed to allow this to take place increases greatly. So, timing of information can influence the participants would have changed or would have not changed their minds or gone home before proceedings could draw to a close. So, price and quality is the main message of information to influence consume individual attitude to decide to buy the product in market.

Timing of information can influence business cycles. It is well known that business activity is cyclinal. The short cycles of 4 to 5 years are best established , but consumers believe who can discern longer term and even very long term cycles of activity with periods of up to 50 years. Cyclical behavior, can only occur where these is an imperfect response to change, because of imperfect information. In general, this sort could not caused by the response time behavior of individual markets. Whatever the nature of the link, it is clearly the case that the price/quality relationship in individual markets is varying independently of the factors which might normally be expected to affect it in isolated systems. So, cyclical phenomena in business are strong evidence of the market process network behaving as a system.

In conclusion, markets are activites to exchange products and services. The elements of economic chains together to allow modern industrial economies or (AI) big data gathering information technological economy to function with all their complexity. They are essential to change and they permit innovation. However, markets are not well represented by the conventional supply/demand schedules, in particular because these can not include the effects of time as an variable factor. It is much clear to represent timing of information as systems in the form of flow diagrams showing the movement of products from innovative processes through markets to consume. Revenue from the market then supplies feedback to product manufacturers, and the whole system responds at different rates to different levels of feedback from clients. An effective medium of exchange is necessary for proper responses to be made. The complexity of the modern world, where price and quantity and quality in the market are all can't exist without the timing of information factor influence. Price signals are often confused by products, imperfect or incomprehensible information and the various effects of time, and in any case quantities to be supplied to the market have to be decided well in advance of market day. The whole trend a modern industrial economy or technological economy is towards product differentiation. Such as mobile phone, laptop computer etc. technologic products. Manufacturers often need to innovate design, quality, functions to adapt to client's individual need. So, technological innovation is often a production of factors to invent high technological products. Services too can't be fitted into price/quantity schedules because it is impossible to define the product. Otherwise, the categorisation of human as labor, having a price/quantity relationship, when who are clearly , each is an individual learning system, changing every day of whose life and changing the economic process accordingly. In fact, human must necessarily be accepted as a feature of modern life, to protect to let them to enjoy high quality of life. So, individual economic stage will be needed to enter technological economic stage in economic environment. Indeed by limiting the rate of change such actions may in no small measure be a condition of stability for the people in an economy . It seems that (AI) big data gathering information technological innovation is one factor of production and it has close relstionship to timing of reasonable price and quality information to persuade consumers to choose to buy the manufacturer's product.

6 (AI) big data gathering information can reduce the cost basis of economic activity to any businesses

(AI) big data gathering information technology can help any organizations to reduce the time and human effort economic cost. In conversion processes there is always some wear and tear of the fixed asset, the equipment, building etc. which reduce their capacity to produce in future. Of course after the money has been spent on the plant, it is no longer cost of operating. This is not technological obsolescence which results from development of better ways of meeting market needs. The efficiency of produrers is continually improved and the most effective use of the resources available is continually improved and the most effective use of the resources available to the society, and the most effective use of the resources available to the society is made according to the criteria of economic values. The under-utilised resources locked up in the inefficient operation are not necessarily lost. The producer may learn in time to use them better. So that who can eventually compete on moral equal terms, or who may give way to another who knows how to manage the resources more efficiently. If however, the inefficiency has a deep-rooted cause which can not be remedied, or even a producer who is capable of improvement but refuses to act, then the resources run down to extiaction faster than ordinary wear and tear would cause them to. So, it suppose to technological innovation can help producers to reduce cost for long term. It is cost benefit to producers for long term.

For agricultural industry example, it was said that the only way for a farmer to increase whose not revenue significantly, once who was farming as efficiently as possible, was to increase the area of land under cultivation. But technological innovation is such production of factor, it might be a case for increasing the area under cultivation in order to make better use of a piece of equipment, such as a tractor, and so spread its cost over more production. Another might be in the processing industries, such as petrochemicals or oil where many new producers with the same global threats and opportunities. The input costs when products or utilities move in the direction of time and energy in the economic process. If one opportunity for using the resources is selected, then another potential use will be foregone. Opportunity costs are therefore distinguished from input costs by time. So, the production for factor of technological innovation can be opportunity cost, if the manufacturer felt who can spend less manufacturing expenditure for long term. Due to who lose to use money to spend other expenditure or invest, who choose to invent technological innovation to reduce long term input costs. The best opportunities are those which maximise prices and minimise costs. SO, (AI) big data gathering technology can be applied to argicultural industry to help farmers to reduce farming time and farming equipment cost to grow any fruit, vegetable and rise , tomotato , potato etc. food in production of factor view.

Producers try to achieve higher market prices by giving their products some distinguishing feature which whose hope will attract buyers away from other competitors and /or generate new buyers, what marketers call product differentiation. In effect they try to move their product into a new market, perhaps thought of as " up market" or a " market niche", but certainly in separate market for analytical purpose. Even if they can not do this, which is unusual in these days of increasing technological innovation and communication , operators continually try to improve their processes in order to reduce costs. If always requires the investment of new resources, e.g. technological innovation.

In the real world, it is not possible to differentiate products or processes except in time. The overall result is to move the process in the direction of economic improvements, in effect the behavior of economic system as a learning system. Time introduces all the risks and opportunities which present themselves to the processor. In the analysis which follows , we classify and illustrate the various aspects of economy of scope under the traditional economic heading of labor, capital, and land. Energy and information, technological innovation are considered too ,because which are fundamental to all systems and process in economic activity.

7 The (AI) big data gathering technological innovation benefits

First, (AI) big data gathering technological innovation can bring to help organizations to reduce staff number and divide labor to raise different department work efficiency and performance benefits. For the division of labor benefits example. This is a complex process into stages in which a worker can specialize, thus allowing who to perform that particular task more efficiently, i.e. at lower cost per unit of effective output than if who had to undertake the whole process. Such an improvement in efficiency results from the more effective learning, greater development of skills and more intensive application over a period of time which becomes possible when a task is easily within the capacity of one person.

It is easily confused with advances in (AI) big data gathering technology, capital investment or scale of operation. Division of complex process into stages may subsequently allow the development of specialised technologies for the individual stages, and this may result in specialised equipment, and hence capital investment . Similiarly, if the process is carried out with less labor and/or lower raw material costs for each unit of output, those concerned may in principle decide either the produce more . Output or the produce the same output with less input. It seems technological innovation can bring low cost benefits. In fact, new technology imposes a diseconomy on the old, eg. functions using old technology are at a cost disadvantage and must adopt or eventually disappear under free competition. So, new technology is the source of growth and adoptation in economy. The solutions to a diseconomy of scape lies in a change of scope, for example, in this case different establishments operating at different times, and perhaps with different prices. If capital investment are differentiated be improved to give longer life and better use, which in effect reduces their cost in use. For example, continuing advantages in technology allow processes to be designed in such a way that which deliver the same output with ever decreasing inputs of materials, labor or energy. Thus waste is minimised by planning and the conservation of process energy, and maintenance is reduced by change of design or the use of new materials. It may often worth spending more on equipment initially to reduce those time dependent costs. This sort of efficiency is the most obvious effect of scope rather than scale.

Deterioration and obsolescence means wear and tear are the changes which occur in artefacts as which are used , i.e. deterioration , or changes in quality or scope with time. These are not simply time effects because which depend both on the original design and on the conditions of use, such as maintenance skills and even simple care and attention. Obsolescence is difference to depreciation, it relates to the battle in the market place . The networks of markets brings products and therefore all conversion processes into competition for the same revenues. Old products will be not popular because which become harder to sell. Obsolescence, therefore depends not only on time, but also on competition, in the same time of business. It seems technological innovation can avoid obsolescence occurrence to old products to raise which competitive ability to the same markets. However, there was hardly an element in the competitive cost structures of conversion processes which was not disturbed in a way which differentiated country form country, industry from industry and firm from firm, such as (AI) big data technological innovation to any old products, which production of factor cost structures is the same basically.

8 What is the relationship between the process of (AI) big data gathering technological innovation and the production of factor?

The term "innovation" is used to describe the deliberate process by which a new product or process comes to be sold in the market. Technology innovation can be applied to conversion processes, which requires the use of energy, or their products, which have an economic energy content. It is therefore a function of all the forces which shape markets: manufacturing, processing, technology, buying, selling, information, prices costs etc. So an innovating organization may be a whole company or it may be an individual. Other forms of innovation (production of factor) relate to the sale of services within what defines as the facilitiation system. That sort of exchange is not specifically considered to be one production of factor, because it involves different adoptation processes and response times, and doesn't of itself add to the quantity of products or utilities sold.

However, invention can not be defined to one production of factor and it is to be distinguished to innovation. We can describe invention is as the process of discovering something completely new, i.e. a new fact or relationship. It is an important scientific advance. Invention enlarges the scope of man's awareness, but it doesn't necessarily have direct economic value in itself. If it is sold, it is the sale of an idea, an exchange of a little creativity for an incentive within the facilities system. It isn't marketed as a new product of a conversion process. No energy of conversion is involved. The great majority of inventions do not enter into the economic process and which do not become innovations until that happens.

If the change of scope results in a new (AI) manufacturing technological process for making a product or utility which is already being sold, this can't be differentiated from the existing product or utility in the market concerned. To be successful the new process must make it at lower unit cost than existing process. The result then is that either the price of the product fulls and processes to improve are imposed to competition or more net revenue is accumulated. So the aims of the factors of technological innovation production include: The introduction of a process for making at a lower unit cost a non-differentiated product which is sold into a commodity market, and the development and sale of a differentiated product which will draw buyers away from other markets, or draw money into the market which would not otherwise have been spent.

The nature of (AI) manufacturing technological innovation how causes factor of production. The process of technological innovation is the arrangement of materials at the elementary, say atomic or molecular level, or of components or the design of new machines, or of the relative positions of components, for example, the location of nodes in networks. There are the three levels at which the scope of the economic process may be changed. However, technological innovation can't seem without some change in the way materials ae ordered. It follows that all technological innovation flows initially from some change in a conversion process. Thus technological in the result of investment in conversion processes, where investment is defined as laying down fixed assets and so it requires a change in the use of energy consumed during conversion to make useful products. The flexible manufacturing system themselves are examples of the third level of innovation, the spatial arrangement of components and hence the link between them. Patterns of communication have changed and are continue to change as a result of new technoloby in the use of energy and the convergence of computer, data maipulation and telecommunication. Flexible manfacturing systems manufacture components in rather than having them made by supplies industries and transported to an assembly plant in batches. The new arrangements reduce both the time of reponse to market changes and all the skills of components which are needed to give flexibility of response in conventional systems , i.e. they give economy of scope.

9 How to response times in (AI) manufacturing technological innovation?

In the terminology which have developed above, the behavioral effects may be considered as adjustment of the scope of the sellers and buyers organizations as the process of acceptance of the innovation in the market proceed. Costs and risks in technological innovation, innovation requires the commitment of resources over a long period, and it is therefore subject to the same kind of risks as any investment in conversion processes. The most obvious risk is that the technological difficulties are in the initial concept, with the result that no returns will be earned and the resources sunk in the investment may have been wasted. There is a set of market-related reasons why technological success may not result in an innovation. By the time, the new process or product is ready for the market, the demand for it may have receded or may never have materialised. This may be the result of fulfiment of the potential users' needs by another technology, i.e. the innovation may be technologically obsolete before it may be because of a change of fashion or styles of living.

Two conclusions may be drawn, firstly, the product manufacturer has a good foresight and understanding is needed when undertaking projects which consume large amount of capital, or it may result in gross waste, because the future is always uncertain, however, the analysis, secondly, there is a limit to the rate of constructive innovation in an economic system, the ratio at which the system can accumulate. Hence, some product manufacturer will feel the technological innovation can be a good production of factor , such as a cost advantage is termed a competitive advantages. It is a broader term than the comparative advantages of traditional economy because of does not depend on a favorable climate or an abundance of natural resource. It is developed and maintained entirely by the skills of the people in the firms which are involved.

Technology is like on the economic process, because once knowledge about transforming inputs into outputs has been obtained, and especially after it has been implemented, it doesn't disappear. Technological innovation moves the whole process. Thus, economy of scope confer permanent advantages on those who have them. Economy of scope is to be obtained from all the elements of the economic process which change with time and these are suspectible to improvement, whether as separate elements. They involve people, and their capacity to learn and improve, and material in all their different forms.

10 Why technological innovation will be one factor of production to technological manufacture industry.

Innovation is the process by which new products' processes methods or services are created. Innovation offers added value for and users by providing better and/or cheaper functionality than previous options. Innovation combines changes in technology, business models, organization etc. The basic idea may be a new technical solutions, a new business model or a change in organization. In a competitive economy, no business can survive long term without updating its products and services or the ways in which are produced or delivered. Innovation policy must promote renewal across all business sectors and not just focus on high technological industries.

Since most innovations are complex and each subsystem has its own limitation , an important part of the innovation process is finding the right balance between conflicting demands. In most cases, there are several possible ways of providing a new function to users, or possible applications of a new technology. Which combination of features the market will prefer can't be predicted with any certainty. Whether the origin was a market opportunity on a new technological capability to one part of the production of factor to the product.

Innovation integrates knowledge from a number of different fields: technology, marketing, design, economic etc. In the production of factor view, it is hard to collect all the necessary competences in a single organization. Because technological products need to be updatd to keep competition in market. Thus, innovation has become a process of constant with suppliers and competitors, with consultants and with academic researchers. In the production of factor view, the capacity to innovate depends on how well different parts of this system are adapted to each other and how well they work together.

Today, the relationship between science and innovation is more complex and interdependent. Science-based technologies, such as microelectronics or biotechnology could not have been developed without scientific understanding, but modern science is equally dependent on advanced technology. Economists tend to prefer technology performance standards, but these risk favouring marginal improvements to existing technologies when discouraging more radical, long term solution. Also, economists tend to think of innovation as a production processes. A more production describes innovations as an experimental learning proces in which organizations and individuals build new competence. This term "research-based competence" is rather than " science-based knowledge or "scientific information".

I shall argue that economists' active process is a better way of think about the relationship between industry and academic research. Whether can the production of factor of technological innovation make use of the tools and results of research in addressing real world problem to manufacture any technological products? This main concern at the time was whether research and innovation were essentially different activities which should be supported in different ways or whether it was important to deal with both aspects together since which were interdependent. However, innovation is a process of searching, experimenting and learning. Consumers can learn about how new products, processes and services are created, how firms build competence for this and what information sources which use. So, I feel technological innovation ought be one part of production process or production of factor to some manufacturers. Such as, searching is needed for better ways of doing worth which things. Experimenting is needed because consumers can't seen in advance the best way of accomplishing a desired outcome or indeed what users really want or need. Learning is needed because actors involved in an innovation process will learn from it. The kind of learning which changes consumers' ability to solve future challenges and opportunities. However, economists often think of innovation is as a production prcocess, where knowledge transformed into a new product. We measure research and development investment, relating these to outcomes in the form of patents, new products and productivity or economic growth. Innovations are not just new technological products or processes, it also mentions organizational innovations, new distribution channels, new business models etc. In fact, it is often misleading to think about technical or organizational innovation as separate processes. Most innovations combine changes in technology, business models organization etc. in production process.

What kinds of product can belong to high technological manufacturing. For example, the world's most advanced steel plants and paper mills can never be classified as high technology because of their complementary need for high levels of investment in fixed capital, and aerospace manufacture is classified as medium technology. The standard definition of high technology measures research and development intensity not the generation or use of advanced technology as such. A far better measure is the proportion of scientists, engineers and highly qualified technicians in the labor force. For computer industry example, innovations in the field of rabotic manufacturing, nanotechnologies and human genetics research all have been enabled by low cost computational and control capabilities supplied by computers and software. Reducing the cost of software important objectives of the U.S. software industry. However, the complexity of the software industry to support the U.S. is computerized economy is increasing at an alarming rate. Software nonperformance and failure are expensive. In actuality many factors contribute to the quality issues facing the software industry. These include marketing strategies, limited liability by software vendors, and decreasing returns to testing.

At the core of these issues is the difficulty in defining and measuring software reliability, usability, efficiency, maintenability and portability. Information problems are further complicated by the fact that even with substantial testing, software developers don't truly know how their products with perform until who encounter real scenarios. The similiar industries with need have technological innovation in the productive process (production of factor), such as automotive and aerospace equipment manufacturers and related electronic communications equipment manufacturers. Quality is defined as attribute factor to different kinds of software product. Defining the attributes of software quality and determining the metrics to access the relative value of each attribute are not formalized processes. Because users place different values on each attribute depending on the product's use, it is important that quality attributes be observable to consumers. The technological innovation is one production of factor to software industry. Due to software attributes have those accurateness, interoperability, security, reliability (maturity, recoverability); usability (understandability, learnability, operability); efficiency (time behavior, resource behavior); maintainability (analyzability, changeability, stability, testability); portability (abaptability, installability, replaceability). It seems that due to software attribute have these characteristics, so it causes technological innovation is one production of factor to software industry.

Nowadays, human needs have been increasing, external factor can influence some industries cause technological innovation is one production of factor need. Together, these tends are going to reshape now human live and work, reorganize our social, economic and political institutions and redistribute power and reward in society. In the longer term, as machine learning and computer power intelligence technological innovation needs from consciousness, as machine learning and computer from consciousness, as improving health technologies allow for biological enhancements and species divergence, and as the final frontier is also needed by space travel, technological and social transformation will increasingly change what is means to be human. However, human have to better understand how our world is changing and by what forces those changes are driven . So, because human have high living quality needs, so it causes many new products have technological innovation to manufacture new product or to raise high quality need to satisfy our daily life. It will cause of factor to some products.

The (AI) manufacturing technological innovation factor can influence the economic of the pork meat production in agricultural industry. For example, the economy of the pork meat production on a farm has been carried out with the help of the method of production functions (factor- product and factor-factor). The influence of the weight of an animal on the daily growth tells us that the growth is increased with the increse of the entry weight to 19kg and with the exit weight of the fattened animal of 100 kg. So, the relationship between the daily growth and the feed costs by a feeding day shows us the tendency of than increase of a daily growth with the increase of the feed costs, e.g. with the increase of labor inputs to 2.6 hours/100 kg of the live weight and the increased profit to 29 monetary units. Labor productivity grows with the increase of the capacity usage to 87% and then it decrease. The economy of agricultural production considerably depends on the development of cattle-breeding as a natural capacity of transforming plant products into high quality cattle products. Cattle-raising production influences the food quality, the development of food production industry, the output of high quality and healthy safe product and the development of agricultural economy. So, it seems technological innovation can be a production of factor to influence farm agricultural industry to assist farmers to apply high, e.g. agricultural technology (skills) produces high quality and tastic of farming met to satisfy consumers' diet needs.

Growth of total factor productivity (TFP) can provide society with an opportunity to increase the welfare of people. In particular, in the simplest framework, change in labor productivity factor depends on change (TFP) and capital deepening. How to change TFP? I shall suppose the technological innovation method is a factor to reduce labor cost, but it can raise labor productivity and products or goods of quality to satisfy consumers' needs in competitive market. Economists often define the knowledge economy as production and services based on knowledge-intensive activities tht contribute to an accelerated pace of technical and scientific technology, as well as rapid absolescence. Knowledge is now recognized as the driver of productivity and economic growth, leading to a new focus on the role of information, technology and learning in economic performance. In the knowledge-based economy, innovation is driven by the interaction of producers and users in the exchange of both codified and tacit knowledge: This interactive model has replaced the traditional linear model of innovation. The knowledge-intensive and high technology, economy tends to be the most dynamic in terms of output and employment growth. Changes in technology and particularly the advent of information technologies are making educated and skilled labor more valuable, and unskilled labor less. So, the technological innovation production of factor will bring skilled labor needs, more and unskilled labor needs less.

Although, it can maximize the benefits of technology for productivity, but it can raise unemployment number of non-skillful labor, because the high technological product firms will choose to dismiss the non-skillful labor and will employ skillful labor when innovation which decide to apply technological innovation method to produce whose products. For example, output and employment are expanding fastest in high technology industries, such as computers, electronic and aerospace. Also, knowledge-intensive service sectors, such as online education, communication and information(long distance call) are growing even faster, such as internet shopping technological business can be production of factor to let universities can teach students from internet, such as distance learning. Internet can be used to adventise and sell products from businessman individual website more easily. Also, mobile can use internet to do same benefits, such as laptop or desktop kinds of high technological computer products. It seems technological benefits can attract consumer individual consumption more easily. So, skillful biased technical change is a shift in the production technology that favors skilled over unskilled labor by increasing its relative productivity and therefore, its relative demand. In fact, skill-biased technical change is a shift in the production technology (factor of production that favors skilled, e.g. more relative productivity) and therefore, its relative demand.

11 How can external and internal factors affect the product and (AI) manufacturing process innovation?

In fact, the competition advantages of a company strongly depends on its possibility to benefit from innovational activities. Understanding the factors how which affect product and process innovation and their effort is necessary to be proved why innovational activities can be the one part production of factors to some new products. It has close relationship between product and business processes innovation and industry maturity and customer needs (demand) technological opportunities and investment attractiveness and company size and export orientation. These external and internal factors can influence innovational activities to some new products.

Nowadays, fast technology development, combined with the globalization and fast changes in with the globalization and fast changes in customer demand, implies that a competitive advantage of a company. So, companies will spans great effort in beating the competition innovations have a vital influence on economic development of a country. On the macro level, innovations have a vital influence on economic development that innovations are more and more present both a developed and developing countries that wish to grow developing countries that wish to grow fast and become developed. If we simply categorize companies all innovative or non-innovative. Among different innovation's categorizations is developed by researchers, the most important are: classification according to the type of innovation to degree of innovativity, innovations can be classified as incremental, semi-radical and radical innovations (Davila et al 2006), who indicates that radical innovations potentially offer huge profits and competitive advantage, but demand considerably high risk level, much company effort need and resource engagement. Otherwise, incremental innovations have more modest returns, but demand lower risk level, level of efforts and resources and are generally more successful. Finally, semi-radical innovations are somewhere between the two of them.

According to (Christensen 2003) explained to an innovations can be sustaining and disruptive. Sustaining innovations can be placed in the whole range from incremental to radical and discuptive are either semi-radical or radical. Sustaining innovations are those that improve existing products or process, disregarding the degree of improvement. Disruptive innovations create a huge growth offering a new of performances which has even it is inferior from the start comparing to existing technologies' performances a potential to become superior. Companies are advised to accept what is the best for their situation and design innovational processes, develop aptitudes, allocate resources and form partnerships in compliance to that decision.

In fact, many external and internal factors can affect companies chose product innovations, because process, innovations or their combination, e.g. factors include industry maturity, customer needs and expectations , technological opportunities, investment attractiveness, intensity of cmpetition, company size, origin of ownership and export orientation. In the industry maturity stage, as a market matures and customer needs become defined in a better way, companies transfer the focus of their competition to expenses and economy of range investing more in business processes in order to make them more effective and more efficient. Customer needs and expectations are essential for process innovations that improve process effectiveness. Orientation to customers and their satisfaction are well-known concept in the field of a total quality management.

The point of view that market demand presents the main determine the rate and activities of an invention because each rational company that tends to make profit is responsive to economic stimuli . According to Schmookler (1962) demand growth is prior to the growth in innovative activites, i.e. market requests guarantee stimuli for companies to innovate and take up new technologies. This concept is popularly called " market pull" in a sense that a market pulls innovations.

12 What is (AI) production of factor knowledge economy ?

I shall give evidences to explain why technological innovation can be one kind of production factor to some technological product manufacture industry nowadays. Nowadays, we are entering the knowledge based economy stage. Knowledge is now recognized as the driver of productivity and economic growth, leading to new focus or the role of information technology and learning in economic performance. The knowledge based economy and its relationship is as traditional economics, as reflected in " new growth theory". Because every technological product manufacturer needs workers to acquire a range of skills and to continuously adapt these skills underlines the " learning economy". The importance of knowledge and technology diffusion requires better understanding of knowledge networks and " national innovation systems".

Firstly, knowledge-based economies which are directly based on the production, distribution and use of knowledge and information. The is reflected in the trend in growth in high technology investment, high technological industries, move highly-skilled labor and associated productivity gains. Also required is tacit knowledge including the skills to use and adapt codified knowledge-based economy, innovation is driven by the interaction of producers and users in the exchange of both codified and tacit knowledge.

Employment in the knowledge-based economy is characterized by increasing demand for more highly skilled workers. The knowledge-intensive and high-technology tend to be the most dynamic in terms of output and employment growth. The science system, essentially public research laboratories and institutes of highest education, carries out key functions in the knowledge-based economy, including knowledge production, innovative technology. So, the traditional functions of producing new knowledge through basic research and educating new generations of scientists and engineers with its newer role of collaborating with industry in the transfer of knowledge and technology. For example, our societies tend to research institutes and academic increasingly have industrial partners for financial as well as innovative purposes, but most combines this with their essential role in more generic research and education.

In general, our understanding of what is happening in the knowledge-based economy is constrained by the extent and quality of the available knowledge-rated indicators. So, available knowledge-rated indicated. So, development of indicators of the knowledge-based economy must start with improvements to more traditional input indicators of research and development expenditures and research personal. Better in indicators are also needed of knowledge stocks and flows, particularly relating to the diffusion of information technologies, in both manufacturing and service sectors; social and privates rates of return to knowledge investments to the impact of innovation technology in productivity and growth.

However, knowledge is such as human being (human capital) and in innovative technology has always been central to economic development. When human is entering the 21 ST century, our output and employment are expanding tastes in high technology industries, such as computers, electronics and aerospace investment is thus being directed to high-technology products and services, particularly information and communicating technologies. Computers and related equipment are the fastest growing component of tangible investment. Equally important are more intangible investments in research and development, the training of the labor force, computer software and technical expertise. Hence, it causes employment is growing in high technology, science-based sectors ranging from computers to pharmaceuticals. Also, research and development causes manufacturing sector is losing jobs. Due to those jobs are more highly skilled and pay higher wages than those in lower technology sectors (e.g. textiles and food processing). Knowledge-based jobs in service sectors are also growing strongly. Indeed, non-production or knowledge workers those who don't engage in the output of physical products, are the employees in most demand in a wide range of activities from computer technicians, through physical therapists to marketing specialists.

Economists continue to search for the foundations of economic growth. Traditional, " production functions" focus on labor, capital , materials and energy, land; however, knowledge and technology are external influence on production. Analytical approaches are being developed. So, that knowledge can be included more directly in production functions. Investment in knowledge can raise productive capacity of the other factors of production as well as transform them into new products and processes from innovative technology.

According to the neo-classical production function, returns diminish is as more capital is added to the economy an effect which may be offset, however, by the flow of new technology. In new growth theory, knowledge can raise the returns on investment, which can contribution to the accumulation of knowledge. Technological change can also raise the relative marginal productivity of capital through education and training of the labor force, investment in research and development and the creation of new managerial structures and worth organization. In fact, incorporating knowledge into standard economic production functions is not easy task, as this factor defies some fundamental economic principles, such as that of scarcity, knowledge is intangible, but labor, capital, land, equipment etc. production of factors which can be tangible or measured. However, some kinds of knowledge can be easily reproduced and distributed at lower cost to abroad set of users, which tends to undermine private ownership. Knowledge is a much broader concept than information, which is generally the " know-what", and "know-why" components of knowledge. There are also the types of knowledge which come closet to being market commodities or economic resources to be fitted into economic production functions.

Knowledge can divide know-why and know-how both kinds of concept. Know-why means to scientific knowledge of the principles and laws of nature. This kind of knowledge underlines technological development and product and process advances in most industries. The production and reproduction of know-why is often organized in specialized organizations, such as research-laboratories and universities. Otherwise, know-how means to skills or the capability to do something. Business judging market prospects for a new product or a personnel manager selecting and training staff have to use their know-how. The same is true for the skilled worker operating complicated machine tools. Finally, knowledge-who becomes increasingly important. Know-who involves information about who knows what and who know how to do what. It involves the make if possible to get access to experts and use their knowledge efficiently. So, knowledge economy brings those conditions to our societies. One hypothesis is that globalization and international competition have led to decrease relative demand for less-skilled workers of the phenomenon; an alterative explanation is that innovative technology change has become more strongly biased in favor of skilled workers, changes in firm behavior is as the main reason for falling real wages for low-skilled workers. Thus, innovative technology and knowledge economy has close relationship to cause knowledge workers can bring high technological products of production of factor in technological product manufacture industry.

13 (AI) Production of factor internal technical skill

Secondly, it is the internal skill biased technical change influences. Skill-biased technical change is a shift in the production technology, that flavors skilled over unskilled labor by increasing its relative productivity and , therefore, its relative demand of innovative technology of production factor. The direction of technical changes i.e. whether new capital complements skilled or unskilled labor may be determined by innovators' economic incentives shaped by relative prices, the size of the market and institutions.

Economic theory views the production technology as a function describing that a collection of factor inputs can be transformed into output, and it defines technical change as a shift in the production of function. In fact, given who observed movements of the production function only concentrates on , such as land supply, labor numbers, equipment supply, capital demand factors. Therefore,To make sense of these recent developments, the concept of factor biased technical change can be another production of factor to influence the new technical products quantities change. For example, the timing of the rise in the skill premium has changed the rapid diffusion of information and communication technologies in the workplace environment in any high technological industry generally nowadays. For example, expenditures in information processing equipment and software, is as a share of U.S. private non-residential fixed investment, rose from 6% in 1960 year to 40% in 2000 year. At the heart of those dynamic change.

This is an improvement in the quality and productivity of all those equipment products, relying heavily on semiconductors like computers, software and switching equipment underlying much of communication technology. In the early adoption phase of a new technology, that those who adapt more quickly can reap some benefits. As time goes by, there will be enough makers learning how to work with the new technology to offset the wage differential. Note the difference with the hypothesis set, where the effect of capital deepening on the skill premium is permanent. Also, information technologies production of factor can reduce costs of data storage, communication, monitoring and supervision activities within the firm which causes a shift towards a new organizational design. In particular, the layers in the hierarchical structure can be reduced, so that the organization of the firm becomes "flatter". So, workers no longer perform routinized, responsible for a wide range of tasks within teams. Therefore, adaptable workers verses at multi-tasking activities benefits is a factor of production to reduce internal cost of any firms.

Due to technological innovation causes the layers in the hierarchical structure can be reduced, so that the organization of the firm becomes "flatter". How technological innovation can influence internal skill biased technical change to orgnizational structure. Development behavioral means it is through managment theory. So, high technological skillful organization will choose to apply theory x more than theory y because this technological innovation will reduce some unskillful staffs and give more effort and duties to those skillful staffs to use high technological skill to do whose jobs daily and who will feel lazy and unhappy to do extra more technological jobs. Theory x assumptions are the average human being dislike of work and work avoid if who can, most people must be controlled, directed or threatened with punishment to adequate effort to action organization objective; otherwise, theory y assumptions are people like to use physical and mental effort to work as natural as play and rest, human being dislike work, a source of satisfaction, threat of punishment are not being effort. Hence, technological innovation can cause organizations to change whose structure and skill staffs need to do more jobs. It causes employer need to give extristic and intrinsic motivations to satisfy whose skillful staffs needs to raise efficiency, e.g. giving more tangible reward, as salary, benefit, security, promotion, good contract of condition of work service, comfortable workplace environment as well as using one ability to achieve who feel apprecation, positive being treating of psychological satisfactory needs.

In technological innovation of organization structure, the management committee needs to concern whose skillful technological labor individual psychological needs. Because technological innovation is one important production of factor and it has close relationship between motivation and staff individual efficiency and productivity. As Maslow's hierarchy of needs indicates people (staffs) mean having satisfied to achieve motivation behavior, the lowest love is basic physiological, the need for food, as salary, safe working condition, then is job security, benefit. Next is friendship at work group, after is promotion, payment increasing, high status of job title. Finally is achievement in work advancement opportunitie creative task in related aspect at work motivation. Hence, after the traditional non-technological innoviation of organization changed the technological innovation of organizational structure, management needs to concern that motivation is needed to develop of behavioral through contributed to management theory. Every organiztion manageer needs to know what its team staffs whose indvidual needs, it includes extrinsic needs, e.g. salary, promotion, security as well as intrinsic needs, e.g. achievement, appreciaton. If the employees feel extrinsic needs are more than intrinsic needs, the managers can consider what extrinsic needs, the managers can consider what extrinsic needs of whose employee individual need. If the employee feels intrinsic needs are more than extrinsic needs, the manage can consider what intrinsic needs the employee individual actual need in order to motivate the skillful worker to work efficiently and raising productivity.

After changing the technological innovation of organizational structure, the management needs to concern how to plan to raise its productivity from its production of factor of technological innovation. Planning is looking ahead, control is looking back, every organization must need strategic plan, operative plan and tastic plan for every department to give aim for its mission objectives. Then it needs to achieve its any short term plans or long term plans efficiently, e.g. how to achieve to produce and to sell 5,000 computers sale objective or how to increase to achieve 20% profit or productivity objective from 10% within one year. So, after the technological innovation production of factor influences the organization needs to find reasons what how to influence it can not achieve these new objectives within one year. Then, it needs to find reasons and revises to solve challenges to control it can achieve it's planning objectives. For example, SWOT method indicates what its internal strengths and weaknesses, external threats and opportunities are. To aim achieve its planning strategic plan every year. Before the organization is not technological innovation, it can't have control is looking ahead, due to planning is looking back because organization can;t know what it's mission and objectives can't achieve to revise if it has no any strategic plan, operational plan and tactical plan for top, middle and low level to let different department managers to know what it's mission and objectives are planned to achieve before the organization has not changed the technological innovation of organizational structure in the year. Besides, after the technological innovation changed the organizational structure. The management needs to concern how to implement it's strategies effectively. The technological innovation of organization needs to change its old long term strategic plan to be new long term strategic plan in the top level, e.g. one year what is its new mission for its technologcial innovation, e.g. Apple brand of computer company needs to innovate its old style computer design to know how to adapt the young client group needs (demand) in this competitive computer technological product industry.

Finally, after innovative organizational structure, mangement needs to concern how to raise skillful labor individual productivity and efficiency, due to who need to increase more effort to do their jobs after technological innovation. I shall indicate on job training method. The advantages on limitations of different approaches to on the job training include the company needs to spend extra time and resources to train staffs or workers to work, when who are on the regular work time. Hence, it will lose staffs to do regular job duties, due who needs to learn how to do their job. So the employer needs to pay higher salary to every job trainer for long term if it needs to train many skillful labor after technological innovation. It can't ensure whether the training employees can work efficiently and know who are not the right staffs to get training. Hence, it will employ the staffs who are not right staffs to accept job training riskly if the mangement have not evaluate who have effort to be train to raise whose productivity and testing personal effort of evaluation is more important to the job trainers.

14 What are the (AI) technical change as exogenous or endogenous production of factor?

Finally, I shall indicate what the change is as exogenous or endogenous factor in the (AI) production function model to cause innovative technology to produce new technological product in manufacture industry. Although, economic theory firstly treated technology change is as a residual, the unexplained part remaining after the contribution of an increased quantity and quality of capital, labor and natural resources in output growth have been accounted for. However, the theory of economic growth reconsidered recently the nature of technological change and the concept of knowledge. Therefore, the new growth economic theory includes research and development is as a factor of influence in the macroeconomic models.

The endogenous or exogenous nature of technological change refers to its source: endogenous is internal to the national economy, being created by domestic private or public enterprise, when exogenous change is external originating from foreign sources. So, it seems research and development workforce is as new factor in the production function model. Although, technological progress, managerial improvements and innovation in general are nowadays largely regarded as key contributors to economic growth. Schumpeter (1939) defines technical progress in terms of production function, which describes the way the production output varies according to the quantity and quality of the input factors. So, the technological change represents the factor that shifts the production function.

From the theoretical viewpoint, it has difficult to separate knowledge from the other factors of the production function. The total labor factor productivity is usually estimated by output the capital and labor factors, weighted by their specific shares. Under perfect competition, the price of the production factors is equal to their marginal productivity, hence, their shares in outputs are equal to their exponents from the production function.

Otherwise, from the empirical point of view, there are difficulties of measurement, especially in the case of value added and research-development variables. So, from all available data on research and development input and outputs, research and development expenditures are most frequently used, along with the number of patents, the technological balance of payments, machinery and tools inputs etc. costs to measure of input in innovation.

Furthermore, the exponents of the new growth theory indicates modeled knowledge is as an output quality of the research and development sector and proved that contrary to the neoclassical conclusions of the diminishing-returns technology, the introduction of the human capital changes the production function into one with increasing returns. Thus, it seems total research and development expenditures are used in this model as a measure of total investments (material and intangible) in the research and development sector. However, in many studies, the research and development stock is calculated as the accumulated value of research and development expenditure after depreciation, a procedure which implies the assumption that all of the research and development expenditure certainly and that it's stock depreciates with a certain fixed rate. Since, long time-series data on R & D are rarely available, other studies assume that the growth rate of R & D expenditure to R & D stock is stable. Hence in innovation technological industry, the labor production factor can be divided into two components total employees population outside the research development sector and the number of employees in research and development. The same types of division was applied to the capital production factor.

Reference

Christensen, C.M. (2003) " the innovator's dilema", Harpercollins, New York.

Davila, T., Epstein, M. J., Shelton.R. (2006) " Making innovation work: How to manage it, measure it and profit from it". Warton school publishing, New Jersey.

Schmookler, J., (1962) " Economic sources of incentive activity, " the journal of economic history. vol. 22
, no. 1 (Mar. 1962), 1-20.

Schumpeter, J., A. (1939), business cycles: A Theoretical Historical And Statistical Analysis Of Capitalist Processes, New York: Macmillan.

Reference

Future of jobs survey, World Economic Forum.

Hauser, J. Tellis, G. J; Griffin, A. 2006. Research On Innovation: A Review And Agenda For Marketing Science. 25(6): 687-717.

J.P. Holdren & P.R. Enrlich, " Human population and the global environment", American Scientist, vol. 62 (1974), pp.282-92.

Joel E. Cohen, How many people can the earth support? (New York: Norton, 1995), pp. 212-36, 261-62.

Mohr, G. J. Griffin, A. 2010. Research On Innovation : A Review And Agenda For Marketing Science. 25 (6): 687-717.

Names of drivers have abbreviated to ensure legibility. Future of jobs survey, World Economic Forum.

" Presence to prosperity", PWC Growth Markets Centre Report: http://www.pwc.com/gx/en/growth-markets-centre/presence-to-profitability.jhtml

FIVE

Creative Technology Improves Social Methods

Reducing global environmental pollution advantages factor

What are the advantages to reduce global environmental pollution to our future societies? If we do not continue to avoid to cause air and water pollution from manufacture or driving car etc. business or enjoyment activities, what negative influences what are caused to our future societies? I shall attempt to explain as below:

Firstly , I shall discuss whether reducing greenhouse gases benefits air quality how and why it can save human future lives. Air quality "co-benefits" result mainly from reductions in air pollutant emissions from the same sources that emit greenhouse gases. For example, replacing a coal-fired power plant with a renewable electricity source, such as wind power, reduces both air pollutant and greenhouse gas emissions. Nowadays, our world has been slow to adopt significant actions to address climate change, as it is a long-term and global problem. The benefits of reducing carbon dioxide today are felt in the future, and since they occur globally, countries may take little action and rely on others to lead. On the other hand, better air quality and improved health are realized rapidly and locally, providing government leaders with tangible benefits from their actions to reduce carbon dioxide. For example, reducing greenhouse gases pollution , it can bring health benefits of Air Pollution Reduction to influence global has more fresh air to let we can breathe, then we can avoid air pollution breath to cause our lung hurt, even death easily. Air pollution is a grave risk to human health that affects nearly everyone in the world and nearly every organ in the body. Fortunately, it is largely a preventable risk. Reducing pollution at its source can have a rapid and substantial impact on health. Within a few weeks, respiratory and irritation symptoms, such as shortness of breath, cough, phlegm, and sore throat, disappear; school absenteeism, clinic visits, hospitalizations, premature births, cardiovascular illness and death, and all-cause mortality decrease significantly. The interventions are cost-effective. Reducing factors causing air pollution and climate change have strong cobenefits. Although regions with high air pollution have the greatest potential for health benefits, health improvements continue to be associated with pollution decreases even below international standards. The large response to and short time needed for benefits of these interventions emphasize the urgency of improving global air quality and the importance of increasing efforts to reduce pollution at local levels.

● The consequences of water and air and water and chemical pollution

Air pollution may bring these effects. Diseases such as amoebiasis, typhoid and hookworm are caused by polluted drinking water.Water polluted by chemicals such as heavy metals, lead, pesticides and hydrocarbon can cause hormonal and reproductive.A polluted beach causes rashes, hepatitis, gastroenteritis, diarrhea, encephalitis, stomach aches and vomiting. Pollution or the introduction of different forms of waste materials in our environment has negative effects to the ecosystem we rely on. How does pollution affect the ecosystem? There are many kinds of pollution, but the ones that have the most impact to us are Air and Water pollution.Pollution or the introduction of different forms of waste materials in our environment has negative effects to the ecosystem we rely on. There are many kinds of pollution, but the ones that have the most impact to us are Air and Water pollution. How does pollution affect humans? Harmful gases and particles in the air come from a range of sources, including exhaust fumes from vehicles, smoke from burning coal or gas, and tobacco smoke. There are ways to limit the effects of air pollution on health, such as avoiding areas with heavy traffic. Thus, air pollutants cause less-direct health effects when they contribute to climate change. Heat waves, extreme weather, food supply disruptions, and other effects related to increased greenhouse gases can have negative impacts on human health.

There are many kinds of pollution, but the ones that have the most impact to us are Air, Water, and chemical pollution. How does pollution affect humans? In the following paragraphs, we will enumerate the consequences of releasing pollutants in the environment. We cause most of the pollution and we will suffer the consequences if we don't stop. We are already seeing its effects in the form of global warming, contaminated seafood, increased cases of lung diseases and more.

We release a variety of chemicals into the atmosphere when we burn the fossil fuels we use every day. We breathe air to live and what we breathe has a direct impact on our health. Over 100 million years of healthy life are lost every single year as a result of air pollution. On average, that's the same as 1 year and 8 months of healthy life lost for every single person on Earth.

Air pollution is the world's 4th most lethal killer

Air pollution is the cause of 8.9 million deaths globally every year. That means that every 4 seconds, someone somewhere on the planet dies from air pollution. The UN has called air pollution the world's worst environmental health risk. Air pollution is also the world's 4th most lethal killer (following malnutrition, unsafe sex, and the lack of safe, clean water and sanitation).

Air pollution from car exhaust affects human reproduction

If pregnant women are exposed to air pollution from car exhaust, it can alter the structure of the chromosomes in the fetus and increase the risks of cancer and various birth defects.

Air pollution and climate change closely linked

The main cause of air pollution as well as climate change (CO2-emissions) is the burning of fossil fuels (oil, coal, gas). A change to greener alternatives such as solar or wind power will therefore help both the climate and human health.

How air pollution influences human health

Breathing polluted air puts you at a higher risk for asthma and other respiratory diseases. When exposed to ground ozone for 6 to 7 hours, scientific evidence show that healthy people's lung function decreased and they suffered from respiratory inflammation.Air pollutants are mostly carcinogens and living in a polluted area can put people at risk of Cancer.
Coughing and wheezing are common symptoms observed on city folks.
Damages the immune system, endocrine and reproductive systems.
High levels of particle pollution have been associated with higher incidents of heart problems.The burning of fossil fuels and the release of carbon

dioxide in the atmosphere are causing the Earth to become warmer. Read about the effects of Global Warming here.The toxic chemicals released into the air settle into plants and water sources. Animals eat the contaminated plants and drink the water. The poison then travels up the food chain – to us.

Water Pollution Effects

Just like the air we breathe, water is vital to our survival. We need clean water to drink, to irrigate our crops and the fish we eat live in the waters. We play in rivers, lakes and streams – we live near bodies of water. It's a precious resource that can easily be polluted and the contamination can be transferred to us and affect our health.

The consumer society is powered by water

Everything we buy, use, eat takes water to produce. Our total use of water through the stuff we buy is represented by "The water footprint". The global water footprint is 9 trillion tons per year or almost 300,000 tons per second.

The consumer society is getting more and more thirsty

Global demand for freshwater is projected to increase 55 % between 2000 and 2050. By 2050, it will reach a massive 5500 square kilometers or 5.5 trillion tons. The main sources for the rise in freshwater use are industry and manufacturing with an expected increase of 400 %. In addition, water demand from electricity-generation will increase 140 % and domestic use 130 %.

The pollution of groundwater resources is increasing

280 billion tons of groundwater is being polluted annually. In 2000, Earth's groundwater resources were being polluted twice as fast as in 1960. Water polluted by chemicals such as heavy metals, lead, pesticides and hydrocarbon can cause hormonal and reproductive problems, damage to the nervous system, liver and kidney damage and cancer – to name a few. Being exposed to mercury causes Parkinson's disease, Alzheimer's, heart disease and death. A polluted beach causes rashes, hepatitis, gastroenteritis, diarrhea, encephalitis, stomach aches and vomiting.Water pollution affects marine life which is one of our food sources. Remember the stories of contaminated shellfish and how those who ate them died?

Plastic pollution

Plastic wasn't invented until the late 1800s and the production of plastic didn't take off until around 1950. But then it really took off. The world has produced over 9 billion tons of plastic since around 1950. 6.3 billion tons (over two thirds!) of this plastic have ended up in the environment - including our oceans. By 2025, there will be a staggering 100 bags of plastic for each foot of coastline in the world! At this point, the ocean will contain around one ton of plastic for every three tons of fish. By 2050, there could be more plastic than fish (by weight) in the world's oceans. Just imagine. Diseases such as amoebiasis, typhoid and hookworm are caused by polluted drinking water.

We live in an ecosystem where the action of one has the potential to affect the many. This can be a good or a bad thing, depending on what the action is. Our mistakes has polluted the environment that we live in and we are waking up and owning to the fact. We are trying to reverse the damage. The good news is that every positive action counts. The small effort you make towards a greener environment can start a healing ripple effect. We may still save what is left of our natural resources and make the world a better place to live in for our future generation.

Chemical pollution

Global production of synthetic chemicals is around 250 billion tons a year. Many of these chemicals find their way into our bodies and the consequences are horrifying. In samples from human beings, a study found as many as 420 different chemicals known to or suspected of causing cancer.Another study found an average of 200 industrial chemicals present in the cord blood of newborn babies.

287 different chemicals were identified in the cord blood.

180 can cause cancer

217 are toxic to the brain and nervous system

208 can cause birth defects or abnormal development.

This is truly terrifying. Especially since the global production of synthetic chemicals is expected to increase six-fold between 2000 and 2050.

On average, we already have around 700 synthetic chemicals in our body that are not a natural part of the human body chemistry. And we know very little about how the combination of these chemicals will affect us.

● Reducing air and water and plastic and chemical pollution different policies implement will be needed to different countries in our future societies.

Over the last decades, energy and pollution control policies combined with structural changes in the economy decoupled emission trends from economic growth, increasingly also in the developing world. It is found that effective implementation of the presently decided national pollution control regulations should allow further economic growth without major deterioration of ambient air quality, but will not be enough to reduce pollution levels in many world regions. A combination of ambitious policies focusing on pollution controls, energy and climate, agricultural production systems and addressing human consumption habits could drastically improve air quality throughout the world. By 2040, mean population exposure to PM2.5 from anthropogenic sources could be reduced by about 75% relative to 2015 and brought well below the WHO guideline in large areas of the world. While the implementation of the proposed technical measures is likely to be technically feasible in the future, the transformative changes of current practices will require strong political will, supported by a full appreciation of the multiple benefits. Improved air quality would avoid a large share of the current 3–9 million cases of premature deaths annually. At the same time, the measures that deliver clean air would also significantly reduce emissions of greenhouse gases and contribute to multiple UN sustainable development goals.

Given the dynamics of these factors and their complex interplay, what could be expected for future air quality around the world, and which determinants will be dominating? To answer this question, this paper identifies key factors that contributed to historic air pollution trends in different world regions, outlines conceivable ranges of their future development and examines their interplay on global air quality in the next decades. In particular, the paper provides a fresh perspective on how ambitious policy interventions could achieve clean air worldwide.

● Future projections of air pollutant emissions

A range of studies in the scientific literature explored the implications of these findings on future emissions and air quality. For a long time, future global air pollutant trends were mainly modelled in the context of long-term greenhouse gas emission scenarios . The early global studies on air pollutant emissions, notably the scenarios developed for the 'Special Report on Emissions Scenarios' and the 'Representative Concentration Pathways' that have been prepared for the Intergovernmental Panel on Climate Change (IPCC) proposed declining trends of (energy-related) air pollutants, due to autonomous technological progress and assumed pollution control policies along the environmental Kuznets hypothesis. Later, the improved understanding of the importance of targeted air quality policy interventions motivated a more differentiated approach to projections of air pollutant emissions, resulting in a wider range of air pollutant trajectories than in previous global scenarios. At the same time, the climate community addressed the interactions between decarbonization strategies and air pollutant emissions, both with the interest to reveal health benefits from low carbon policies and to explore the combined impacts of long-lived greenhouse gases and short-lived air pollutants (e.g. SO2 and black carbon) on radiative forcing and temperature increase . In general, the literature reveals strong impacts of ambitious decarbonization strategies on energy-related air pollutants SO2, NOx and PM, due to the phase-out of fossil fuels and the containment of all flue gases connected with carbon capture and storage. However, enhanced use of biomass as a greenhouse gas policy measure may lead to higher PM emissions . Compared to the climate-focused analyses that deal mainly with energy-related emissions and the role of climate policy interventions, only a few studies addressed the longer-term prospects for air pollution from a health- and ecosystems perspective. These studies take full account of other sources that also contribute substantially to population exposure to harmful air pollution, such as agricultural activities, waste management and materials handling. Also, they developed a more holistic approach towards the understanding of future trends in nitrogen emissions and their health and environmental impacts.

- How air pollution may influence the course of pandemics

The COVID-19 pandemic is causing devastating mortality, with the highest rates of intensive care unit hospitalization and morbidity among older adults, men, and those with certain preexisting conditions, most notably cardiopulmonary diseases, obesity, and diabetes. In addition, a host of interrelated socioeconomic factors—including race, ethnicity, occupation, and poverty—increase the risks of COVID-19 infection for people of color, health care professionals, and other essential workers. These factors are, in turn, influenced by conditions of the human environment including chronic levels of air pollution, most notably fine particulate matter (PM2.5) that is a well-established risk factor for death from cardiovascular and pulmonary obstructive diseases. This raises the question of whether long-term exposure to higher levels of PM2.5 increases the severity of COVID-19 and, if so, what measures might be taken to ameliorate those risks. This is the challenge addressed by Wu et al. in a new contribution to a developing series of papers for Science Advances that is dedicated to the study of pandemics from an environmental perspective.The ideal way to address questions about how PM2.5 pollution might influence the course of the pandemic would involve the study of detailed health datasets for very large numbers of people from all walks of life and locations. In this way, the potential effects of PM2.5 pollution might be evaluated in the context of other details about each individual's life history and conditions. The amount of time required for rigorous, extensive studies, however, conflicts with the swift nature of the COVID-19 pandemic. Addressing the potential impact of air pollution on COVID-19 mortality requires a more nimble approach to environmental policy decision-making.

COVID-19–related death counts (compiled by Johns Hopkins University for more than 3000 U.S. counties) and well-established PM2.5 pollution levels for each county. The results show that higher values of exposure to PM2.5 are positively correlated with higher county-level mortality after taking into account over 20 potentially confounding factors. Most notably, they conclude that an increase of just 1 μg/m3 in the long-term average of pollution is associated with a significant 11% increase in a county's rate of mortality.There are strong policy implications for these results. COVID-19, zoonotic influenza, and other potentially severe emerging zoonotic diseases are and will remain long-term threats to our species. Rapidly emerging datasets suggest that these threats are likely to be exacerbated by air pollution, even at the levels currently attained in the United States despite conscientious efforts to improve air quality. While incomplete and not yet fully vetted by the broader scientific community, pathfinding studies such as that of Wu et al. set the stage for more traditional environmental epidemiology research.

● Benefits of Reducing and Reusing policy

Recucing and reusing policy may help our earth to avoid serious pollution influences , such as prevents pollution caused by reducing the need to harvest new raw materials, saves energy, reduces greenhouse gas emissions that contribute to global climate change, helps sustain the environment for future generations, reduces the amount of waste that will need to be recycled or sent to landfills and incinerators and allows products to be used to their fullest extent.

● Ideas on How to Reduce and Reuse to implement

Buy used. You can find everything from clothes to building materials at specialized reuse centers and consignment shops. Often, used items are less expensive and just as good as new. Look for products that use less packaging.

When manufacturers make their products with less packaging, they use less raw material. This reduces waste and costs. These extra savings can be passed along to the consumer. Buying in bulk, for example, can reduce packaging and save money. Buy reusable over disposable items. Look for items that can be reused; the little things can add up. For example, you can bring your own silverware and cup to work, rather than using disposable items. Maintain and repair products, like clothing, tires and appliances, so that they won't have to be thrown out and replaced as frequently. Borrow, rent or share items that are used infrequently, like party decorations, tools or furniture.

Thus, we are living in our earth. We are everyone has responsibilities to do environmental protection activities in every day, such as reduce and reuse activity will be our right environmental protection daily behavior, walking replaces to reducing to driving when we need short time to arrive the destination in any time, manufacturers need to buy air and water clean machines to avoid serious air and water pollution in their factory manufacturing processes, we need to reduce the frequeny to travel, e.g. one to two times travelling by air planes every year, then sky will have much fresh air in our earth, also airlines need to shorten flying time , e.g. New Zealand airlines only fly to Australia near distance country , it can not fly to US, or UK far away distance countries, China airlines only fly to Singapore, Japan etc. near distance Asia countries, they do not fly to US, UK far away disrance countries. Then, our future environment pollution will be reduced as well as we can have much fresh air to breathe and drive clean water to proplong our lives when we have health.

Improving internet technology development factor

Why do we improve to improve internet technology? What long term social benefits will benefits if scientists can improve internet speed and reseach any information function ? I shall research these questions to give suggestion as below:

Why does internet improvement make life better? Internet of Things Benefits In short, the scale of change that IoT technology offers can be scary. At the same time, the benefits of a well-executed IoT strategy can be more need for an organization: Safety, Comfort, Efficiency. Also, the Internet offers teens the ability to make friends with peers with whom they would not otherwise connect. With pop culture deteriorating into many distinct subcultures, teens' interests are more variable than they have ever been.With internet communication, employees can effortlessly communicate with one another at anytime from anywhere in the world. This allows employees situated in different parts of the world to give their opinion and voice their concerns. Through internet access, individuals in developing countries are able to gain access to more of the modern economy. With internet connectivity, those living in remote areas can now easily take out microloans, participate in e-banking and more. A large share of respondents predict enormous potential for improved quality of life over the next 50 years for most individuals thanks to internet connectivity, although many said the benefits of a wired world are not likely to be evenly distributed.

- How internet can excite young to learn?

Internet can learn youngs to learn much different new knowlege when they research any questions and find answers from internet channel.

As one major aspect of teen life is social environment, changes in how teens connect impact the ways in which teens develop social skills. ** Luckily, the Internet offers many social-skill enhancement opportunities for teens of all different personalities . One advantage the Internet brings that the standard school environment cannot is the ability for teens to adjust their amount of social interaction. Teens who are extremely outgoing can spend their free time in social environments both offline and online, making new connections and catching up with friends.For example, a teen who finds large amounts of face-to-face interaction to be intimidating can use the Internet to engage in conversations while reducing the potential for social anxiety. In a way, this trains less social teens to be more social . In the past, these types of teens did not have the advantage of this social training provided by the Internet.

- Internet can encourage Social Network Growth

The Internet offers teens the ability to make friends with peers with whom they would not otherwise connect. With pop culture deteriorating into many distinct subcultures, teens' interests are more variable than they have ever been. Whereas in the past, children at school might have discussed the current top 40 when discussing music, today's kids define their musical tastes as specific genres, such as post-industrial, dubstep or jpop. Today, it's harder for teens to find peers who share the same interests in their schools. But online, not so. The Internet's social networks help teens find communities of peers who share similar interests, allowing a teen to grow his social network in a way that is specific to him 2. Today's teens are increasingly willing to make friends with different groups of people due to the ability to actually meet them, and this can be useful when they reach adulthood, a time in which accepting people of different backgrounds and demographics is crucial to career and academic growth. The Internet offers teens the ability to make friends with peers with whom they would not otherwise connect.

Today's teens are increasingly willing to make friends with different groups of people due to the ability to actually meet them, and this can be useful when they reach adulthood, a time in which accepting people of different backgrounds and demographics is crucial to career and academic growth.But the Internet can help teens foster self identity through exposure to new people, communities, hobbies and concepts. As teens go through more experiences, they learn more about themselves. And as the Internet can offer teens a wealth of experience, it can play the role of hastening the development of self identity.For many teens, the hardest part of life is figuring out identity.But the Internet can help teens foster self identity through exposure to new people, communities, hobbies and concepts.

● What Are Main Benefits of Internet Communication speed improvement ?

It may include as below:

1 Makes communication easier

Doing business through phone or mail doesn't work well ? Before the internet came into existence, the only way to communicate was through a phone. Or if you needed to send a note you had to send letters via mail. With the arrival of the Internet, staff and team managers can connect instantaneously without leaving their work place. ezTalks Meetings, a one-stop internet communication provider, is a perfect example. With this platform, participants can communicate as if they were right next to one another thanks to its quality video and audio. The tool comes with a rich set of features like screen sharing, cross platform chat, innovative whiteboard, and more.

2 Enhances collaboration

Internet communication brings teams together across the globe. Staff can collaborate easily without limitations and make more informed decisions instantaneously. This leads to reduced project timelines, cutting back on the time required to launch a new product/service. This piece of technology is also useful in education. Not only can students collaborate with foreign students, they can share ideas and learn about the diverse cultures out there. Parents can also become actively involved in their kids education by linking their children school with libraries, homes, and more. Millions of schools around the world are already using this technology to enhance learning.

3 It is cost effective

The cost of internet communication is significantly low when compared with other means of communication like face to face meetings and mail delivery. The technology connects you to your partners, colleagues, clients and suppliers from just about any location for a fraction of the cost required to host a one-on-one meeting. And as technology continues to become more efficient, the cost of online communication continues to drop significantly. With the traditional face to face meeting, you need to spare time, cash to travel and so on. Internet communication allows you and your team to connect without having to leave your offices.

4 Improves work relationships

Building a good relationship between workers spread around the globe is not easy. Business trips can negatively affect life– work balance. Team members can burn out fast if they have to make business travels that deny them the chance to participate in crucial events with friends and family. With internet communication, employees can effortlessly communicate with one another at anytime from anywhere in the world. This allows employees situated in different parts of the world to give their opinion and voice their concerns. Therefore, internet communication is an important business asset, particularly for companies that have tapped into global markets.

5 Increases productivity

While the companies of yesteryear might not have treasured effective communication, modern workplace requires both the management and the staff have the tools to effectively communicate internally and externally. This is because effective communication is important in increasing productivity as it directly impacts the behavior of the employees and how they perform. Internet communication plays an integral role in getting stuff done fast and efficiently which ultimately improves productivity. Poor communication can have a negative effect on productivity as the staff may not get the adequate info to accomplish a job they have been assigned.

6 Increases accountability

Errors slow down productivity and so it is tempting to punish or fire employees who repeatedly make errors. One major advantage of internet communication is that it helps to decrease these errors. This piece of technology pinpoints errors and how staff can avoid them. In workplaces that don't make use of various forms of internet communication, those mistakes go unnoticed. With internet communication, there is no room for mistakes as employees feel liable for their actions and safe to point out mistakes. They also feel secure expressing their ideas and suggestions in a group setting.

- Why does internet improvement can help any industries services or efficiencies improvment?

Internet improvement will revolutionize the world and lead to groundbreaking changes in transportation, industry, communication, education, energy, health care, communication, entertainment, government, warfare and even basic research. For example, self-driving cars, trains, semi-trucks, ships and airplanes will mean that goods and people can be transported farther, faster and with less energy and with massively fewer vehicles. Automated mining and manufacturing will further reduce the need for human workers to engage in rote work. Machine language translation will finally close the language barrier, while digital tutors, teachers and personal assistants with human qualities will make everything from learning new subjects to booking salon appointments faster and easier. For businesses, automated secretaries, salespeople, waiters, waitress, baristas and customer support personnel will lead to cost savings, efficiency gains and improved customer experiences. Socially, individuals will be able to find AI pets, friends and even therapists who can provide the love and emotional support that many people so desperately want. Entertainment will become far more interactive, as immersive AI experiences come to supplement traditional passive forms of media. Energy generation and health care will vastly improve with the addition of powerful AI tools that can take a systems-level view of operations and locate opportunities to gain efficiencies in design and operation. AI-driven robotics (e.g., drones) will revolutionize warfare. Finally, intelligent AI will contribute immensely to basic research and likely begin to create scientific discoveries of its own. So, it implies that internet improvement ought assist any kinds of industy service or efficiency improvement.

- Internet may become any organizational digital assets

On an individual basis, we will think about our digital assets as much as our physical ones. Ideally, we will have more transparent control over our data, and the ability to understand where it resides and exchange it for value – negotiating with the platform companies that are now in a winner-take-all position. Some children born today are named with search engine-optimization in mind; we'll be thinking more comprehensively about a set of rights and responsibilities of personal data that children are born with. Governments will have a higher level of regulation and protection of individual data. On an individual level, there will be greater integration of technology with our physical selves. For example, I can see devices that augment hearing and vision, and that enable greater access to data through our physical selves. Hard for me to picture what that looks like, but 50 years is a lot of time to figure it out. On a societal level, AI will have affected many jobs. Not only the truck drivers and the factory workers, but professions that have been largely unassailable – law, medicine – will have gone through a painful transformation. It seems entirely reasonable that a great deal of our digital lives will be focused on habitable environments: identifying them, improving them, expanding them.

Significant, often highly communication and computation technologically driven, advances in day-to-day areas like health care, safety and human services, will continue to have a significant measurable improvement in many lives, often 'invisible' as an unnoticed reduction in bad outcomes, will continue to reduce the incidence of human-scale disasters. Advances in opportunities for self-actualisation through education, community and creative work will continue. So, I believe that future many organizations may apply internet communication tool for their digital assets.

- Internet improvement may assist robotic development

Most of the focus on technology and particularly AI and machine learning developments these days is limited to virtual systems (e.g., apps for travel booking, social networks, search engines, games). I expect this to move, in the next 50 years, into networking people with machines, remotely operating in a myriad of environments, such as homes, hospitals, factories, sport arenas and so on. This will change work as we know it today, as it will change medicine (increasing remote surgery), travel (autonomous and remotely-guided cars, trains, planes), entertainment (games where real robots, instead of virtual agents, evolve in real scenarios). These are just a few ideas/scenarios. Many more, difficult to anticipate today, will appear.

They will bring further challenges on privacy, security and safety, which everyone should be closely watching and monitoring. Beyond current discussions on privacy problems concerning 'virtual world' apps, we need to consider that 'real world' apps may enhance many of those problems, as they interact physically and/or in proximity with humans. So, future historians will observe that, in many ways, the rise of the internet over the next few decades will have improved the world, but it hasn't been without its costs that were sometimes severe and disruptive to entire industries and nations as well as improve robotic development.

This is similarly valid for AI.Living longer and better lives is the shining promise of the digital age. Many respondents to this canvassing agreed that internet advancement is likely to lead to better human-health outcomes, although perhaps not for everyone. As the following comments show, experts foresee new cures for chronic illnesses, rapid advancement in biotechnology and expanded access to care thanks to the development of better telehealth systems. Life will improve in multiple ways. One in particular I think worth mentioning will be improvements in health care in three distinct ways. One is significantly better medical technology related to cancer and other major diseases. The second is significantly reduced cost of health care. The third is much higher and broader availability of high-quality health care, thereby reducing the differences in outcomes between wealthy and poor citizens. So, when hospitals can improve internet communication , if the hospital can apply robots to assist doctors and nurses to serve patients. Then, internet communication can help them to cooperate more efficient.

- Internet improvement to assist 5G laptop development

Many of the technologies we see commercialized today began in government and university research labs. Fifty years ago, computers were the size of walk-in closets, and the notion of personal computers was laughable to most people. Today we're facing another shift, from personal and mobile to ambient computing. We're also seeing a huge amount of research in the areas of prosthetics, neuroscience and other technologies intended to translate brain activity into physical form. All discussion of transhumanism aside, there are very real current and future applications for technology 'implants' and prosthetics that will be able to aid mobility, memory, even intelligence, and other physical and neurological functions. And, as nearly always happens, the technology is far ahead of our understanding of the human implications. Will these technologies be available to all, or just to a privileged class? What happens to the data? Will it be 'willed' as a digital legacy to future generations? What are the ethical (and for some, religious and spiritual) implications of changing the human body with technology? In many ways, these are not new questions. We've used technology to augment the physical form since the first caveman picked up a walking stick. But the key here will be to focus as much (or more) on the way we use these technologies as we do on inventing them. All of above factors will be influenced to future 5G mobile phone by internet improvement?

Our homes, transportation, appliances, communication devices and even our clothes will be constantly communicating as part of a digital network. We have enough pieces of this today that we can somewhat imagine what it will be like. Through our clothes, doctors can monitor in real time our vital signs, metabolic condition and markers relevant to specific diseases. Parents will have real-time information about young children. The difference in the future will be the constant sharing of information, data updates and responses of all these interconnected devices. The things we create will interact with us to protect us. Our notions of privacy and even liability will be redefined. Lowering the cost and increasing the effectiveness of health care will require sharing information about how our bodies are functioning. Those who opt out may have to accept palliative hospice care over active treatment. Not keeping track of children real-time may be considered a form of child neglect. Digital will do more than connect our things to each other – it will invade our bodies. Advances in prosthetics, replacement organs and implants will turn our bodies into digital devices. This will create a host of new issues, including defining 'human' and where the line exists between that human and the digital universe – if people are always connected, always on are humans now part of the internet?

● How internet improvement influences AI provides medical service to hospitals?

Similarly, AI embedded in devices or wearables can be applied to predict and ameliorate many mental health illnesses. However, there is potential for there to be huge inequalities in our societies in the ability of individuals to access such technologies, causing both social disruption and new causes for mental health diseases, such as depression and anxiety. On balance, I am an optimist about the ability of human beings to adjust and develop new ethical norms for dealing with such issues.Surveillance technology, especially that powered by AI algorithms, is becoming more powerful and all-present than ever before. But to look at that and say that technology won't help people is absurd. Medical technology, technology to help people with disabilities, technology that will increase our comfort and abilities as humans will continue to appear and develop.The digital revolution will bring benefits in particular for health, providing personalized monitoring through Internet of Things and wearable devices. The AI will analyze those data in order to provide personalized medicine solutions.The most noticeable change for better in the next 50 years will be in health and average life expectancy. At this pace, and, taking into account the developments in digital technologies, I hope that several discoveries will reduce the risk of death, such as cancer or even death by road accident. New drugs could be developed, increasing the active work age and possibility maintaining the sustainability of countries' social health care and retirement funds. Another area AI can have impact is in creating the framework within genomics, epigenomics and metabolomics can be used to keep people healthy and to intervene when we start to deviate from health. Indeed, with AI we may be able to hack the brain and other secreting cells so that we can auto-generate lifesaving medicines, block unwanted biological processes (e.g., cancer), and coupled to understanding the brain, be able to hack at neurological disorders."

Thus, I believe that future hospitals were able to utilize internet technology to solve human health problems to make citizens' lives better and improve their access to care and services to improve their health outcomes. The benefits of the internet in the health care industry have continued to improve access to care and services, particularly for elderly, disabled or rural citizens. Digital tools will continue to be integrated into daily life to help the most vulnerable and isolated who need services, care and support. With laws supporting these groups, benefits in these areas will continue and expand to include behavioral health and resources for this group and for others. In the area of behavioral health in particular, digital tools will provide far-reaching benefits to citizens who need services but do not access them directly in person. Access to behavioral health will increase significantly in the next 50 years as a result of more enhanced and widely available digital tools made available to practitioners for delivering care to vulnerable populations, and by minimizing the stigma of accessing this type of care in person. It is a more affordable, personalized and continuous way of providing this type of care that is also more likely to attain adherence.

- The cyborg generation: Humans will partner more directly with technology when internet is popular to be used in any where

The inevitable 'Singularity' will result in changes to humans and will increase the rate of our evolution toward hybrid 'machines.' I also believe that new and modified materials will become 'smart.' For instance, new materials will be 'self-aware' and will be able to communicate problems in order to avoid failure. Ultimately, these materials will become 'self-healing' and will be able to harness raw materials to manufacture replacement parts in situ. All these materials, and the things built with them will participate in the connected world. We will see continued blurring of the line between 'real' and 'virtual' life." For exaple, artificial general intelligence and quantum computing available in a future version of the cloud connected to individual brain augmentation could make us augmented geniuses, inventing our daily lives in a self-actualization economy as the conscious-technology civilization evolves. Implants in humans that continuously connect them to the web will lead to a loss of privacy and the potential for thought control, decline in autonomy.

- Everyone agrees that the world will be putting AI to work, when internet is improvement to raise robotic efficiency and performance improvement

The technology visionaries surveyed described a much different work environment from the current one. They say remote work arrangements are likely to be the rule, rather than the exception, and virtual assistants will handle many of the mundane and unpleasant tasks currently performed by humans. The shooting is done by a drone guided by a smart guy/gal working a 9-to-5 job in an air-conditioned office in a nice town. Garbage could be picked up, sorted, recycled, all by robots with AI. Tedious surgery completed by robots and teaching via YouTube would leave the humans to the interesting and exciting cases, not the redoing of same lessons to yet more patients/students. Humans could live well on a 20-hour work week with many weeks of paid vacation. Having a job/career could become a positive, not just a necessity. With 24/7 learning and just-in-time capacity, people could change areas or careers many times with ease whenever they become bored. This positive outcome is possible if we collectively manage the creation and distribution of the tools and access to the use of new emerging tools. Thus, future everyone will have hundreds of digital workers working for them. Our cognitive mediators will know us in some ways better than we know ourselves. Better episodic memories and large numbers of digital workers will allow expanded entrepreneurship, lifelong learning and focus on transformation.

Thus, our future social development already small world will shrink further as remote collaboration becomes the norm, resulting in major social changes, among them allowing the recent concentration of expertise in major cities to relax and reducing the relevance of national borders. Furthermore, deep learning and AI-assisted technologies for software development and verification, combined with more abstract primitives for executing software in the cloud, will enable even those not trained as software engineers to precisely describe and solve complex problems. I believe the question we're facing is not 'When will machines surpass human intelligence?' but instead 'How can humans work together with machines in new ways?' Rather than worrying about an impending Singularity, I propose the concept of Multiplicity: where diverse combinations of people and machines work together to solve problems and innovate. In analogy with the 1910 High School Movement that was spurred by advances in farm automation, I propose a 'Multiplicity Movement' to evolve the way we learn to emphasize the uniquely human skills that AI and robots cannot replicate: creativity, curiosity, imagination, empathy, human communication, diversity and innovation. AI systems can provide universal access to sophisticated adaptive testing and exercises to discover the unique strengths of each student and to help each student amplify his or her strengths. AI systems could support continuous learning for students of all ages and abilities. Rather than discouraging the human workers of the world with threats of an impending Singularity, let's focus on Multiplicity where advances in AI and robots can inspire us to think deeply about the kind of work we really want to do, how we can change the way we learn and how we might embrace diversity to create myriad new partnerships. So, future AI and internet technoloy will become new partners to assist any business development, even any organizations and social development. Hence, internet improvement must be needed in order to let any businesses can apply robots to raise efficiencies and improve performance more effectively. For example, free internet-connected devices will be available to the poor in exchange for carrying around a sensor that records traffic speed, environmental quality, detailed usage logs, and video and audio recordings (depending on state law). There will be secure vote-by-internet capabilities, through credit card or passport verification, with other secure kiosks available at public facilities (police stations, libraries, fire stations and post offices, should those continue to exist in their current form). Internet and 24/7 real-time connectivity will no longer be viewed as a 'thing' independent from daily life, but integral, like electricity. This has profound psychological implications about what people assume as normal and establishes baseline expectations for access, response times and personalization of functions and information. Contrary to many concerns, as technology becomes more sophisticated, it will ultimately support the primary human drives of social connectedness and agency. As we have seen with social media, first adoption is noncritical – it is a shiny penny for exploration. Then people start making judgments about the value-add based on their own goals and technology companies adapt by designing for more value to the user . Technology is going to change whether we like it or not – expecting it to be worse for individuals means that we look for what's wrong. Expecting it to be better means we look for the strengths and what works and work toward that goal. Technology gives individuals more control – a fundamental human need and a prerequisite to participatory citizenship and collective agency. The danger is that we are so distracted by technology that we forget that digital life is an extension of the offline world and demands the same critical, moral and ethical thinking.

In future 50 years every aspect of our life will be connected, organized and hence, partly controlled, as technology platform and applications businesses will take this opportunity. A few global players will dominate the business; smaller companies (startups) will mostly have a chance in the development sector. Many institutions, such as libraries, will disappear – there might be one or two libraries that function as museums to show how it used to be. People who experienced today's world will definitely value the benefits and amenities they have through technology (human-machine/AI collaboration). If technology becomes part of every aspect of our lives we will have to give up some power and control. People thinking in today's terms will lose a certain amount of freedom, independency and control over their lives. People born after 2030 will probably just think these technologies produced changes that are mostly for the better. It has always been like this – people have always thought/said 'in the old days everything was better. The free, open internet that represented a set of decentralized connections between idiosyncratic actors will be recognized as an aberration in the history of the internet. Today's internet giants will probably be the internet giants of 50 years from now. In recent years, they've made substantial progress in curtailing innovation through acquisitions and copying. As the industry matures, they will add regulatory capture to their skill sets. For many people around the world, the internet will be a set of narrow portals where they exchange their data for a curtailed set of communication, information and consumer services. Thus, digital tools will be part of our body inside and remotely, and will assist us in decision-making constantly, so it will become second nature. Nonetheless, physical feelings will still be exclusively 'physical,' i.e., there will be a significant difference between the 'sensor-based feelings' and real body feelings, so human beings will still have some advantages over technology. This, I believe, will last forever.

Discovery new health medicine drugs factor

Why do we need to concern new health medicine drugs discovery? I believe that human will face any new kinds of diseases that we had not encountered or contacts in my past. If we lack any new kinds of health medicine drugs discovery to fight any kinds of new diseases in my future. Then, we must face death very easily, such as COVID 19 is one kind of new disease, the another person or other persons can be contacted to cause this kind of COVID 19 disease by the patient's cloths, shoes, hands, even air, mouth of hs body and things. Thus, it had caused many people die in global nowadays. So, medicine or drug or bio-scientists need to spend much time to do any experiment to attempt to discover any new kinds of medicines or drugs to fight any future new kinds of diseases. Otherwise, human will die very easily in soon.

- Why do we need drug discovery?

In the fields of medicine, biotechnology and pharmacology, drug discovery is the process by which new candidate medications are discovered. Historically, drugs were discovered by identifying the active ingredient from traditional remedies or by serendipitous discovery, as with penicillin. More recently, chemical libraries of synthetic small molecules, natural products or extracts were screened in intact cells or whole organisms to identify substances that had a desirable therapeutic effect in a process known as classical pharmacology. After sequencing of the human genome allowed rapid cloning and synthesis of large quantities of purified proteins, it has become common practice to use high throughput screening of large compounds libraries against isolated biological targets which are hypothesized to be disease-modifying in a process known as reverse pharmacology. Hits from these screens are then tested in cells and then in animals for efficacy.

However, modern drug discovery involves the identification of screening hits, medicinal chemistry and optimization of those hits to increase the affinity, selectivity (to reduce the potential of side effects), efficacy/potency, metabolic stability (to increase the half-life), and oral bioavailability. Once a compound that fulfills all of these requirements has been identified, the process of drug development can continue. If successful, clinical trials are developed. Modern drug discovery is thus usually a capital-intensive process that involves large investments by pharmaceutical industry corporations as well as national governments (who provide grants and loan guarantees). Despite advances in technology and understanding of biological systems, drug discovery is still a lengthy, "expensive, difficult, and inefficient process" with low rate of new therapeutic discovery. For example, in 2010, the research and development cost of each new molecular entity was about US$1.8 billion In the 21st century, basic discovery research is funded primarily by governments and by philanthropic organizations, while late-stage development is funded primarily by pharmaceutical companies or venture capitalists. However, discovering drugs that may be a commercial success, or a public health success, involves a complex interaction between investors, industry, academia, patent laws, regulatory exclusivity, marketing and the need to balance secrecy with communication. Meanwhile, for disorders whose rarity means that no large commercial success or public health effect can be expected, the orphan drug funding process ensures that people who experience those disorders can have some hope of pharmacotherapeutic advances.

● Where do new drugs come from? Why does it take so long to get a new drug approved? Why are drugs so expensive?

The medicines we ingest, inject, and inhale are often complex therapeutic compounds. The drugs are usually mixtures of chemicals made from starting materials or drug sources. Depending on the sources from which the drugs were created, the drugs can be categorized as natural, synthetic, or semi-synthetic. Natural drugs are made from compounds found in nature. The most prevalent natural drug sources are plants. The field of science that studies the relationship between people and medicinal plants is known as medicinal ethnobotany. Some examples of medicine that come from plants are morphine (from opium), digoxin (from flower, Digitalis lanata), and aspirin (from willow tree bark). Less prevalent natural drug sources include animals, microbes, and minerals. The first kind drug source is for example, synthetic drugs come from starting materials that are not found in nature. Instead, they are produced by man from smaller chemical building blocks. An example of synthetic medicine is the experimental anti-malaria drug, arterolane. Another kind drug source is semi-synthetic drugs are neither completely natural nor completely synthetic. They are a hybrid. Semi-synthetic drugs are generally made by converting starting materials from natural sources into final products via chemical reactions. Examples of semi-synthetic medicine include the antibiotic, penicillin, and the chemotherapy drug, paclitaxel. To make the chemotherapy drug, paclitaxel, 10-deacetylbaccatin is extracted from yew needles and undergoes a 4-stage synthesis process. They both are the main kinds of drugs manufacturing sources.

● Why does human need new drugs discovery ?

The reason of global health needs demand new approach to drug discovery, the pharmaceutical industry has made enormous strides in the production of potential therapies and medicines. But even today, close to 90% of candidate drugs that enter Phase 1 trials fail to make it to the market place. This is a system beset by duplication of effort and hence wastage of resources. No one lab or institution can do this on its own. We must urgently pool resources and expertise, minimise duplication, explore new drug targets, biomarkers, and technologies in order to generate new, effective, and more affordable drugs for patients more quickly.

Discovey of any one kind of new drug, it needs long time to experiment. It must come up with new ways to accelerate our drug discovery process. Alternatively, we must entirely rethink how we treat illness. This is not just limited to bacterial infections. We need to invent better ways to combat all forms of disease. The process of discovering, testing, and approving a drug for commercial use can take 20 years and over of 1 billion dollars. Obviously, decreasing both the time and the cost of developing these drugs can save many lives. There are some new technologies which are already helping to ramp up this process. For example, computational modeling of drugs has massively sped up the screening process for drugs. We can now take thousands of potential drug candidates and narrow them down to a couple viable options. But there are more ways we can expedite this process.

A recent estimate states that we now know the molecular cause of over 4,000 diseases — but we only have drugs for about 250. How can we do better? The FDA approval process is long and arduous. Even for compounds that have been approved in other countries, FDA trials can be drawn out for years. The FDA approval process can be responsible for about 25% of the cost of a drug and can delay the arrival of a drug over 10 years. There is even data that suggests that the FDA kills many more by not approving drugs than it ever saves by approving drugs (for more on the harmful effects of the FDA, see Cato, Forbes, The Independent Institute, and LifeExtension). By delaying good drugs that can save lives, and by doing little to stop bad drugs, the FDA is often an inhibitor to the medical process. We need to rethink the FDA if we want to streamline the drug discovery process. If we can change many FDA policies, we will see more drugs created for those 4,000 known targets.

● The process of new drug experiment success time evaluation

Any new kind of drugs experiment success, they must experience these processes. They may include:

1 Drug testing and licensing

All new drugs and treatments have to be thoroughly tested before they are licensed and available for patients. A new drug is first studied in the laboratory. If it looks promising, it is carefully studied in people. If trials show that it works well and doesn't cause too many side effects, it may be licensed. You may hear this process called 'from bench to bedside. There is no typical length of time it takes for a drug to be tested and approved. It might take 10 to 15 years or more to complete all 3 phases of clinical trials before the licensing stage. But this time span varies a lot. There are many factors that affect how long it takes for a drug to be licensed.

2 Factors that affect how long trials take

The type of cancer drug success experiement needs time

Clinical trials for rarer cancers often take longer because there are fewer patients available to take part. Research teams from several different countries may need to collaborate so there are enough patients. This can mean the trial takes longer to organise and set up. But international trials can often recruit people more quickly and so are likely be quicker in the long run.

Researchers running clinical trials for more common cancers are generally able to find enough people to take part more easily.

3 The type of treatment

Trials that use new methods of giving treatment, such as a new way to give radiotherapy for example, may take longer to set up and run. This is because the research teams need specialist equipment and extra training. These trials may only be able to run in a small number of hospitals compared to trials using standard ways of giving treatment. How long treatment takes can also affect the results. It is likely to be quicker to get results for a trial looking at a single dose or short course of treatment, compared to a treatment that lasts for months or even years.

4 The type of trial

Some trials look at treatments to prevent cancer or ways of screening for cancer. Screening means testing for cancer in people who don't have any signs or symptoms. People who join these trials haven't been diagnosed with cancer. The research team will often want to follow them for many years to see who develops cancer and who doesn't. They will then compare the different trial groups to see if a particular treatment can help prevent cancer or whether a test can help to diagnose it early.These trials often take a long time to get results compared to treatment trials. It can take years to see a clear difference in the number of people in the different groups who go on to develop cancer. So, any new kinds of drug research experiements, they depend on the number of patients needed in order to decide whether how drug quality level, how many drugs manufacturing supply number, drug price in global market.

Statistics experts look at what the research team want to find out and the design of the trial, and then work out how many patients are needed. If there aren't enough patients taking part, the results may not be reliable. The number of people they need to get reliable results will depend on how many treatment groups there are and exactly what the research team want to find out.

5 The follow up period

Research teams look at how well people are doing for some time after they have treatment as part of a trial. This is to see how well the treatment works over a longer period of time, and to find out more about long term side effects. Follow up periods can range from a few months to more than 10 years, depending on the type of treatment and the group of patients. Or maybe longer for a trial looking at screening or prevention.

6 Any problems with the new treatment

There may be problems with new drugs or treatments that the researchers don't know about until they run the trials. There could be unexpected side effects or reactions to treatment. Or there may be difficulties in giving the treatment to patients. Problems with the new treatment may mean the trial takes longer to complete.

Thus, any kinds of new drug experiement need long time to be attempted to carry on, every new kind of drug experiment is evaluated about 10 to 20 , even more time. So, future drug scientists have responsibilities to evaluate whether which kinds of diseases will cause in order to concentrate on spending time to carry on researching the kind of new drug experiment. It aims to use limited resource and time to let patients to get health.

Improving living environment factor

What are the disadvantages if we do not concern how to improve our global living environment? I shall explain as below:

● reasons to improve living environment

Nowadays, as population on the earth keeps expanding, human needs increase endlessly causing more global environmental problems to proliferate globally. Global environmental crisis has become an unequivocal fact that can affect our livelihood and it is capable of changing the current landscape drastically. Hence, people hold the responsibility to tackle current global environmental issues to make this world a better place. With destructive natural disasters like flash floods or snowstorm as well as the changing of weather patterns, the earth is poised at the precarious verge of severe environmental crisis. Human intervention has caused many dysfunctions to the environment, some of which have left damages on the ecosystem that eliminates other sources of necessity to other living things. Ever since humans start to harvest the Earth 's resources, many landscapes have been altered to fit the lifestyles of countless inhabitants. So people ought to be aware of other types of environmental challenges that the planet is facing. Some of the challenges that the planet is facing include overpopulation of human beings that leads to natural resources depletion, deforestation and loss of biodiversity, acid rain and ocean acidification, pollution and waste disposal. Thus, it seems that global living environment and pollution have close relationship. I mean that air and water and paste and chemical pollution will reduce if we can keep our global living environment more clean, safe and without more rubblishs are allowed to keep in our living places, even gardens, public places anywhere.

One of the key ethical questions is whether a life-extension pill would extend our healthy years or simply prolong frailty towards the end of life. Better health and longer life would certainly be an attractive prospect for many people. If we were healthier for longer then perhaps we could achieve more of our ambitions and engage in the things we enjoy for longer. But some people worry that our lives may be extended in a state of low quality of life rather than health. Although this is not the goal, critics worry that it might be an unintended consequence of intervention in ageing and longevity. As with all pharmaceuticals, both health benefits and risks need to be considered. If life span could be extended a great deal – perhaps to more than 100 years or even longer – then some other interesting issues might arise. For instance, would we simply run out of things to do and become bored? Even things that we enjoy may become stale after several centuries. How long would we have to work for? If our lives were 200 years long then it is unlikely that many people could afford to retire at 65. However, this may also present new opportunities such as having several different careers within a lifetime. If our future earth can not provide a health and clean living environment to let human to live, then our quality of living must be worse to compare nowadays, it will cause our next generation can not be health to live ot they will have many different kinds of disease, such as COV19 disease , or future there are many kinds of serious disease to compare COV19 disease , they will bring threats to influence our next generation to live in anywhere health places in our earth. It is very disappointment to us, such as our next generation's parents, we have not feel responsibilities to keep our living environment to be improved to let our next generation to live in global anywhere clean and health living environment. So, we need to concern how to improve our living environment nowadays.

However, I shall suggest these methods how to improve our living environment to be better. If we can be habit to do environment protection behaviors every day, then our living environment must be influenced to improve more easily and rapidly, in society, individual sand businessmen and our governments have resposibilities, they may include as below:

● Individual and businessmen and governments how to improve living environment

Individual responsibilities to improving living environment

1. Use Reusable Bags

Plastic grocery-type bags that get thrown out end up in landfills or in other parts of the environment. These can suffocate animals who get stuck in them or may mistake them for food. Also, it takes a while for the bags to decompose. Whether you are shopping for food, clothes or books, use a reusable bag. This cuts down on litter and prevents animals from getting a hold of them. There are even some stores (such as Target) that offer discounts for using reusable bags! These bags are useful for things other than shopping as well. I have heard of people using reusable bags when they move! If you forget your bags at home, buy a new one. Better yet, keep a couple bags in your car so you never leave home without them (just make sure you remember you put them there)! If you are in a position where you need to use the plastic bags, reuse them the next time you go shopping, or use them for something else. Just do not be so quick to throw them out!

There are some states that are outlawing or charging extra for using plastic bags. Using reusable bags helps the environment AND your budget!

2. Print as Little as Necessary

We have all had that teacher that wanted us to have a copy of every single reading when we come to class, or that professor who wanted a hard copy of the ten-page paper that is due next week. These are fine but it seems as if they do not understand that using so much paper is detrimental to the environment. What can you do? Ask your teacher if you can bring a laptop or an e-reader to class so that you can download the reading onto that and read it from there. If not, print on both sides of the page to reduce the amount of paper used. If you need to turn in a long paper, ask the professor if it is okay to print on both sides of the page and explain why you're asking. Most teachers care about the environment as well and would be willing to allow you to do so.

3. Recycle

Recycling is such a simple thing to do, but so many people don't do it. Many garbage disposal companies offer recycling services, so check with the company you use to see if they can help you get started! It is as simple as getting a bin and putting it out with your trash cans for free! Another way to recycle is to look for recycling cans near trashcans. Instead of throwing recyclables in the trash with your non-recyclables, make a point to take an extra step to locate recycling cans around your campus.

4. Use a Reusable Beverage Containers

Instead of buying individually-packaged drinks, consider buying a bulk container of the beverage you want and buying a reusable water bottle. Not only will this help the environment, but it will also help you save money since you are buying a bulk container. Many campuses offer water fountains designed for drinking as well as for refilling reusable water bottles. Make use of these fountains throughout the day when you finish off the initial beverage.

Along these lines, many restaurants offer reusable containers for drinks. If you go to a certain place a lot, consider buying one of these containers to help minimize waste. A lot of coffee shops even offer a discount to customers who use a reusable container for their drinks. Starbucks, as an example, offers a small discount for customers who do this. Saving the environment and money?

5. Save Water

Water is wasted more frequently than we can see. Turn off the faucet as you are brushing your teeth. Don't turn your shower on until you're ready to get in and wash your hair. Limit your water usage as you wash dishes. Changing old habits will be good for both the environment and your wallet!

6. Avoid Taking Cars or Carpool When Possible

Cars are harmful to the environment. Taking public transportation, walking, or riding a bike to class are better options that help the environment and your budget, as well as getting some exercise in! If you do need to use your car, compare schedules and places of residency with those in your classes. You can split the cost of gas and have alternating schedules for who drives when. This is cheaper than everyone driving separately and you'll be closer with friends!

Businessmen responsibilities to improving living environment

Instead of individuals have respobsibilities to improve our living environment. Businessmen have also responsibilities to improve our living environment. I shall indicate mining businessmen example to explain how mining businesses may influence our living environment to be worse. The disadvantages of mining include harm to air pollution, water pollution, loss of usable land, destruction of animal habitat, and harm to local communities and the miners themselves. While mining produces the resources needed for fuel, electronics, and other items as well as jobs, companies often don't factor the harm mining can do into their decision making. Below factors may influence our global living environment to be worse as below:

Air Pollution

Lead, arsenic, cadmium, and other harmful substance As are often exposed by mining and picked up by the wind, causing allergies and breathing problems in local people. Mining machinery uses fossil fuels and releases large amounts of carbon dioxide and other substances that contribute to global warming.

Water Pollution

Mining can cause metal contamination and acid mine drainage that makes water unsafe for plants and animals. Sediments released by mining choke streams and erode soil. Both of these problems also cause problems for farming and the water people drink.

Loss of Usable Land

Mining, especially open pit mining, destroys land that can be used for farming, houses, and other human purposes, often permanently. Entire mountains and rivers can be destroyed. Loss of soil and deep underground excavation can also make land unstable and collapse.

Destruction of Animal Habitats

Mining also has disadvantages for plants and wildlife. It destroys homes and food sources for animals and leads to less diverse plant and animal life. Endangered species that are already sensitive to changes in their environment are especially at risk. Because mining releases toxins that linger for years later, the damage to plants and animals often isn't fully understood until after mining has ended.

Harm to Miners

Mining is dangerous for the people who do it, especially for miners who work underground. Breathing in mineral dust can cause deadly diseases like pneumoconiosis or black lung, while the machinery used often causes hearing loss. Back injuries and other physical problems are also common in miners. While big disasters often show up in the news, many of the miners who are killed or injured on the job never receive media attention. In 2010, almost 2,500 miners died from causes other than major accidents.

Consequences for Local Communities

Mining is also harmful to the communities that support mines. Mining can lead to loss of homes, land, and clean water, and it often releases chemicals into the environment that cause health problems for locals. Mines also need large amounts of water to operate, which leaves less for people to drink or farm with. It also causes less obvious problems. Because only some people in an area benefit from mining, but everyone faces at least some of the disadvantages, mining can divide communities. It can also lead to harassment or abuse from corporate or government officials who care more about the profits of mining than the people it affects. The secrecy around mining and who makes money from it often makes this disadvantage even worse.

Thus, ourselves and businessmen can not neglect our any activities can influence global living environment to be worse. We need to learn how to avoid to do any bad behaviors to influence our future global living environment to be worse, even the worst.

Governments responsibilities to improving living environment

Any country's government needs to concern social responsibility before it decides to implement any sustainable development. Because although sustainable development may bring some benefits to some countries, but it can also bring disadvantages to themselves countries. What Are Disadvantages of Sustainable Development? It may brins these disadvantages as below:

Increased Costs

Because sustainable development relies on newer technologies and materials that cost more to produce, the overall costs are often more than that of traditional construction. The higher cost of materials is passed on to developers. Developers pass it on to property owners, who pass it on to tenants. Future development will include tools that haven't even been invented yet. The trial and error of using new materials and ideas can also bring costs up for everyone.

Lower Quality of Life for Some Elements of Society

Sustainable development will shrink or do away with certain job sectors. This will lead to job loss for some workers. The fossil fuel industry could see plants close and employees lose jobs as sustainable development relies on new energy sources. The rising costs and less robust power of alternative energy can also lead to a lower quality of life for people who live in sustainably developed areas.

Resistance to New Methods

When people try to implement new ideas, there's naturally a certain amount of resistance. People in general are set in their ways and don't want to change their lives radically. As more governments and companies attempt to put sustainable development into practice, more resistance will follow. Some of the resistance will come in the form of people who initially adopt the idea of sustainable development with enthusiasm, but their commitment shrinks as they start to put new ideas into practice. Contractors and tenants may resist a specific initiative because it forces them to change the ways they work and live.

Increased Regulation

Sustainable approaches will naturally lead to increased regulation on construction and the daily operation of businesses. A greater commitment to the environment will lead to tighter controls on how people live their lives. Stricter building codes and tougher emission standards are likely. While some people will accept a greater burden of regulation because they see the overall benefit, many people will disagree with government intruding into their lives.

Political Struggle

In addition to the public resistance, there's a political cost to committing to the environment. The deep political divides in society mean that some political powers won't want to commit to sustainable initiatives. Certain industries will try to influence politicians via lobbyists. Some politicians will be completely against sustainable development.

Is It Worth the Trouble?

People and organizations that are in favor of sustainable development believe that it's worth moving past these disadvantages to work on the environment. Advocates say that sustainable development is an investment in future generations. The biggest defenders of these initiatives are working on ways of overcoming the hurdles.

Thus, ineffective or poor sustainable development may also bring poor living environment, due to wrong sustainable development to the country. For example, if Afria government only concern how to find mining lands for sustinable development, but it neglects to keep clean and health and natural land living environment to African to continue to live. Then, it will reduce African quality of living to be worse. So, any countries governments need to keep balance to bring social benefit when they decide to do sustainable development in themselves countries.

Improving social welfare factor

Do you feel our nowadays social welfare is enough? What factors can influence our social welfare to be better or worse? Do we need to improve our social welfare to let our next generations feel comfortable and without difficulty to live? What feeling will influence to our future next generation if we do not plan to improve our social welfare? I shall explain why we need to concern how to improve our social welfare in order to let our next generation won't difficult to live as below:

● Why do we improve future social welfare ?

Why do we need to improve social welfare? Firstly, I shall explain why we need social welfare. Then, I shall explain why we need to improve social welfare. Social welfare may be explained that it is one kind of social protection has the potential to reduce insecurity for workers and help to bring employment contracts. Also, it is an investment in human capital for economic growth as successful economic depend on the quality of their workforce.

Why do we need social welfare? As a social welfare system offers assistance to individuals and families in need with such program as health care assistance, food providing and unemployment compensation. Lesser knon pasts of a social welfare system include disaster reief and educational assistances. What is the importance of welfare? When the welfare state has played aim important role to any countries in reducing socio-economic inequalities and proetcting people from various forms of handship , such as unemployment and ill health , as also proverty be an important social problem for economic development.

What is the purpose of social welfare policies? Social wefare policies mean providing especially assistance and social insurance benefit, traditionally have been conceived as instruments of social protection and redistribution. At a minimum, social welfare policies should protect individuals form proverty and relative deprivations So, it brings this question: who benefits from social welfare? The most common types of programs provide benefits to the elderly or retired, the sick or invalid, dependent survivors, mothers, the unemployed, the work-injured, and families. Methods of financing and administration and the scope of coverage and benefits vary widely among countries may be implemented in popular.

Nowadays, our societies believe that social welfare is an important tool for redistribution, social protection which has to be at the heart of the construction of the European project. If social and labour market policies are conceived in an appropriate manner, they help to promote both social justice and economic efficiency and productivity, instead of providing education, medical retired etc. welfare to the low income, low educational level social group as well as to achieve reducing poverty and inequality aim. Thus, it explains why our societies need to learn how to improve our global future social welfare to be be better , even the best social welfare system.

On conclusion, why was social welfare needed to create? Our society's population had been increasing rapidly and human's age had been prolonging, due to medical improvement, enough food provision. However, global has may dependent children and poor old people would gradually need as employment improved, retired assistance, unemployment assistance, educational assistance, and those over 65 age began to collect social security pensions. So, if we can not improve our societies to have the best social welfare to assist, these the most need of social assistance group. Consequently, the difference of rich and poor group must increase . It will be unfair to those low education and old age and low income families group in global. Thus, our future society must need to implement safe feeling to let anyone feel ourselves countries are suitable to us to live for our next generation. Hence, learning how to improve social welfare must be need to every country nowadays.

● Can improve social welfare to influence economy growth?

How social welfare impact any country itself economy? Does it has relationship between social welfare and economy? When it comes to public discourse the term "welfare state" is most often used in a derogatory way. For many people who hear the term welfare state, it means money being handed to people in poverty who don't deserve it because they aren't working to earn their income. The prevailing logic is that everyone knows hard workers are rewarded with higher income. However, that term means something different to actual economists who have dedicated their lives to studying economic systems and their impacts on broader society. In fact, the welfare state doesn't only apply to allocating resources to those living in poverty. It also means allocating resources to corporations. Any time the government allocates resources to any recipient in society, it is considered part of the welfare state. Capitalism lends itself naturally to economic cycles. The economy tends to swing severely between booms and busts. Without any kind of social insurance though, a capitalist economy may not recover from the bust end of a cycle. Even if it does, it would take much longer to recover than it would with social insurances in place. It is not in the best interest of the economy or society for a bust to last too long. The health of the economy is dependent on the economic health of the members of society.

However, Any type of government intervention is viewed as against a pure capitalist system. However, capitalism on paper has not worked out as well in practice without some government intervention on behalf of the greater good of society. Sometimes this has looked like a low level of resource distribution to those less fortunate and sometimes it has taken the form of resource distribution to corporations. Even an example such as farmers getting subsidies is a form of the welfare state. The guiding principle of welfare economics should be bringing all shareholders of the economy to a state of equilibrium where all groups share in the feeling of economic well-being.

However, the equilibrium doesn't happen all by itself. It requires public policy through government regulation and intervention to guide the economy in the direction of widespread well-being without sacrificing growth. Welfare economics can't end the bust end of the capitalist economic cycle. What it is meant to do, though, is mitigate the negative impacts of economic recessions. Welfare economics is meant to ensure that the bust end of the cycle isn't too severe and doesn't last too long. We should not be looking to cause undue economic suffering for any members of society.

I beleive that social welfare can impact economy growth. It means that the country can have better economy growth improvement, if it can provide good welfare to itself country. Otherwise, the country can have worse economy growth, if it can not provide good welfare to itself country. The reaons are becaure welfare can include any corporate (company) it's income as well as social individual both. There are two major types of welfare in the welfare state: social welfare and corporate welfare. While there are people in both types of welfare that do take advantage of the system, both are important tactics for stabilizing the economy.

1. Social Welfare

Social welfare encompasses programs such as social security, Medicare/ Medicaid, food stamps, unemployment, the Affordable Care Act and other similar programs. The idea of these programs is to help safeguard people from poverty. The vast majority of the people who need to use these programs are either children, or they've worked their entire lives in our economic system. Fraud is extremely rare in these programs.

2. Corporate Welfare

Corporate welfare comes in a few different forms as well and is very much a type of welfare state. It is tax money that is given to corporations. It can also come in the form of tax cuts to corporations and it can also look like subsidies in certain industries. The purposes of using corporate welfare are usually to help grow a certain industry, to help stabilize a certain industry, or to help a certain industry avoid financial ruin. An example of corporate welfare is the auto-industry bailout. It is believed that corporate welfare actions such as these prevent an even worse economic disaster. Sometimes though, it seems corporate welfare moves make little sense. An example is the most recent Trump tax cuts. The economy has been growing for the past 8 years and is still doing well. Injecting government funds to booming industries through a corporate welfare action like the Trump tax cuts does not seem to fit any of the usual categories for stimulating the economy or preventing a downturn. It is unclear what the effects will be, but the move has been widely controversial among economists.

Hence, if the country has good social welfare system to provide itself country any company sale increasing chance. Then, when the country has many companies can earn high income, when many customers buy their products or consume their lesiure services. Consequently, the country's economy may be influenced to grow rapidly. The corporate welfare side of the welfare state can have some benefits. For instance, the automobile industry bailout saved jobs and a possible worse recession. By the government investing in budding renewable energy programs, it can generate economic growth in areas with innovative solutions to head off an energy crisis. Corporate welfare money doesn't always have that same effect on the economy. There should be greater scrutiny over how corporate welfare is used and whether it contributes significantly to economic growth. All corporate welfare programs should provide benefits to the whole economy.

On the other side, countries such as Norway have found that social welfare boosts their economies and capitalism. Economics professors in Norway have discovered that despite the short-term sacrifice of providing higher wages for their upper-class citizens, in the long term their social welfare programs have resulted in better equality, smaller gender wage gaps, and improved education across the nation—just to name a few examples given by Science Nordic.

Do welfare states boost economic growth ? If in the future human labour is less needed, keeping societies stitched together may require us to reinvent the welfare state. The laws of economics say social welfare should be in accordance with the economic development level of a country. Welfare programs that are beyond a country's development level are not good for economic development, as has happened in Greece. On the other hand, if the economy develops rapidly without corresponding improvement in people's living standards and public welfare, people will not feel a "sense of gain", which in turn will have a negative impact on economic development.

First, excessive welfare beyond a country's development level will impede accumulation and harm welfare programs in the future. In economics, production is the top priority and it decides consumption. A society has to improve its production level if it wants to improve its consumption level. Production here refers to extended production, because only expanding the scale will breed competition and provide unfailing supply. The expansion of scale should be high-quality and high-level expansion of production through innovation and improvement of the industrial structure. Second, welfare at any level needs economic support. High levels of welfare in countries such as Sweden depend on high taxation and high deficit. But the high-level welfare in Greece depends on high debt. High welfare supported by high taxation reduces development funds for enterprises, impeding the development of enterprises. And if enterprises lose energy, the entire economy will suffer. High taxation also affects individuals' desire and capacity for consumption and thus undermines people's enthusiasm to expand production. Third, excessive welfare will breed dependence and result in waste of social resources. Although high welfare comes from individual taxpayers' contribution, it seems like a public welfare provided by the state. It will result in many social problems, such as waste of social resources, voluntarily unemployment and retirement in advance. Once people get used to this kind of dependence, economic development will be undermined. economic development will also be undermined if the authorities fail to provide enough welfare for the people.

For China economy development example, China social welfare and itself economy deveopment has exact close relationship. There is a lesson to be learned here from the planned economy to China economy growth and its social welfare can be improved nowadays. China's social welfare level today is not high; there is much room for improvement. So to strike the right balance between welfare and economic development, we should abide by the following principles:

One, it has to be clarified that the basic and final goal of China's economic development is the well-being of the Chinese people. And since China is the world's second-largest economy, it should pay more attention to improving public welfare. The Fifth Plenum of the 18th Communist Party of China Central Committee said the national GDP and urban and rural residents' incomes have to be doubled by 2020 compared with the 2010 level, and hence the authorities should focus on coordinated development to improve public services.

Two, the distribution of public welfare should be fair and transparent. The public welfare different social groups enjoy today is unbalanced, especially when it comes to urban and rural areas. Therefore, the authorities should make efforts to rectify the imbalance.

Three, the authorities should take measures to prevent unfairness and corruption from creeping into redistribution of welfare.

And four, they should not forget that China is still a developing country, and development is key to solving social economic problems, and only further development can guarantee sustainable and high-level welfare. More importantly, development problems should not be used as an excuse to reduce public welfare.

On conclusion, I believe that improvment to any country itself social welfare, it can impact the country itself economy growth or recession significantly. Thus, any country needs to concern how to improve itself social welfare in order to improve economy growth, instead of improvement social poor problems for our future generation.

World peace factor

One important issue that human must need to concern if we hope our future societies can develop or improve in success. This issue is that " our world must need to keep peace. Why and how our word must need to keep peace ? The reason is very simple. If our future countries aim to only to be the world top leader, every leader only concentrates on finding the best method to become world top leader, he/she neglects to concern how to make himself/herself country to develop technology, medicine, construction, social welfare etc. different social aspect issues. His/her time only concentrates on war aspect, so peace world is the major key to help our next generation can have safe, clean and good economy social environment to work and live in our world anywhere. The question is : How to keep peace world to avoid war? Ambitious leader must not respresent that he/she must be the top leader that he/she has effort to dominate any countries matter. He/she ought need to know peace is the most important factor that when every country can cooperate to do any matter in order to solve any social challenges, due to cooperation can help any country leader to reduce time to find the best solution when any county leader faces difficulties. So, leader cooperation must need to keep world peace in our future society, moreover, any country improvement must need any country leader cooperation to sit down to discuss any important issues together. It is the best method to build global the most safe society in order to let any country citizen to feel comfortable and safe to live in themselves countries. I shall explain how we can keep peace world as below:

Why do we need to achieve world peace? The reason most scoff at the notion of achieving world peace is because if you buy the principle that individual human revolution is the real solution, then literally some billions of people would need to actively embrace the notion of devoting themselves to continual self-reformation.Peace, as we all know, is very important in our lives and it is essential to our overall well-being. However, this is something that has, regrettably, eluded us for years and years in this world.We will be able to lead others into Peace, rather than getting drawn into their conflict and the endless cycle of action and reaction that only leads to more and more violence. There is increasing evidence to suggest that being peaceful, for example through yoga, tai chi or focused meditation, can have a direct and measurable impact on the world.It matters because we have an opportunity to break the cycles of violence, fear, and hatred- to help one another to wake up. By keeping our hearts and minds open we model how not to walk blindly down paths that lead towards only more hate and revenge. Be the peace you want to see in the world.

On conclusion, future we need to concern how to apply creative technology to above these aspects in order to improve our future societies to be more safe as well as reduce war occurrence as well as improve our qualify of living.

www.ingramcontent.com/pod-product-compliance
Ingram Content Group UK Ltd.
Pitfield, Milton Keynes, MK11 3LW, UK
UKHW041856190726
13854UKWH00002B/937

9 798887 045214